Springer Proceedings in Physics

Volume 293

Indexed by Scopus

The series Springer Proceedings in Physics, founded in 1984, is devoted to timely reports of state-of-the-art developments in physics and related sciences. Typically based on material presented at conferences, workshops and similar scientific meetings, volumes published in this series will constitute a comprehensive up to date source of reference on a field or subfield of relevance in contemporary physics. Proposals must include the following:

- Name, place and date of the scientific meeting
- A link to the committees (local organization, international advisors etc.)
- Scientific description of the meeting
- List of invited/plenary speakers
- An estimate of the planned proceedings book parameters (number of pages/articles, requested number of bulk copies, submission deadline).

Please contact:

For Americas and Europe: Dr. Zachary Evenson; zachary.evenson@springer.com
For Asia, Australia and New Zealand: Dr. Loyola DSilva; loyola.dsilva@springer.com

Abhishek Tewari · Nikhil Dhawan ·
Gautam Agarwal · Sourav Das · Sumeet Mishra ·
Anish Karmakar
Editors

Proceedings of the 3rd International Conference on Advances in Materials Processing: Challenges and Opportunities

AMPCO 2022, 17–19 Oct, Roorkee, India

Springer

Editors
Abhishek Tewari
Department of Metallurgical and Materials Engineering
Indian Institute of Technology Roorkee
Roorkee, Uttarakhand, India

Nikhil Dhawan
Department of Metallurgical and Materials Engineering
Indian Institute of Technology Roorkee
Roorkee, Uttarakhand, India

Gautam Agarwal
Department of Metallurgical and Materials Engineering
Indian Institute of Technology Roorkee
Roorkee, Uttarakhand, India

Sourav Das
Department of Metallurgical and Materials Engineering
Indian Institute of Technology Roorkee
Roorkee, Uttarakhand, India

Sumeet Mishra
Department of Metallurgical and Materials Engineering
Indian Institute of Technology Roorkee
Roorkee, Uttarakhand, India

Anish Karmakar
Department of Metallurgical and Materials Engineering
Indian Institute of Technology Roorkee
Roorkee, Uttarakhand, India

ISSN 0930-8989 ISSN 1867-4941 (electronic)
Springer Proceedings in Physics
ISBN 978-981-99-1970-3 ISBN 978-981-99-1971-0 (eBook)
https://doi.org/10.1007/978-981-99-1971-0

This Springer imprint is published by the registered company Springer Nature Singapore Pte Ltd.
The registered company address is: 152 Beach Road, #21-01/04 Gateway East, Singapore 189721, Singapore

Contents

Contributors

Dinesh Agrawal Materials Research Institute, Pennsylvania State University, Pennsylvania, USA

Munmun Agrawal Department of Metallurgical Engineering, Indian Institute of Technology (Banaras Hindu University), Varanasi, India

Ramkishor Anant Department of Materials and Metallurgical Engineering, Maulana Azad National Institute of Technology, Bhopal, MP, India

A. Anderson School of Mechanical Engineering, Sathyabama Institute of Science and Technology, Chennai, India

Botcha Appalanaidu GITAM (Deemed to be University), Visakhapatnam, Andhra Pradesh, India

Markush Bakhla Materials and Metallurgical Engineering Department, NIAMT, Hatia, Ranchi, India

G. K. Bansal Materials Engineering Division, CSIR-National Metallurgical Laboratory, Jamshedpur, India

Jayashree Baral Department of Materials and Metallurgical Engineering MANIT Bhopal, Bhopal, India

Shamik Basak Department of Mechanical and Industrial Engineering, Indian Institute of Technology Roorkee, Roorkee, India

Jayanta Bhattacharya School of Environmental Science and Engineering, Indian Institute of Technology, Kharagpur, West Bengal, India;
Department of Metallurgical and Materials Engineering, Indian Institute of Technology, Kharagpur, West Bengal, India

Maitreyee Bhattacharya Metal Extraction and Recycling Division, CSIR-National Metallurgical Laboratory(CSIR-NML), Jamshedpur, India

A. K. Chandan Materials Engineering Division, CSIR-National Metallurgical Laboratory, Jamshedpur, India

B. Chandrasekhar Nuclear Fuel Complex, Hyderabad, India

Kausik Chattopadhyay Department of Metallurgical Engineering, Indian Institute of Technology, Banaras Hindu University (IIT-BHU), Varanasi, India

S. Chenna Krishna Materials Processing Division, Vikram Sarabhai Space Centre, Trivandrum, India

S. Ghosh Chowdhury Materials Engineering Division, CSIR-National Metallurgical Laboratory, Jamshedpur, India

Sandip Ghosh Chowdhury Academy of Scientific and Innovative Research (AcSIR), Ghaziabad, India;
CSIR-National Metallurgical Laboratory (CSIR-NML), Jamshedpur, India

G. Das Nuclear Fuel Complex, Hyderabad, India

Ghanshyam Das Department of Materials and Metallurgical Engineering, National Institute of Advanced Manufacturing Technology, Ranchi, India

N. S. Dubey Homi Bhabha National Institute, Mumbai, India;
Nuclear Fuel Complex, Hyderabad, India

Akshay Dvivedi Department of Mechanical and Industrial Engineering, Indian Institute of Technology Roorkee, Roorkee, Uttarakhand, India

Dheerendra Kumar Dwivedi Department of Mechanical and Industrial Engineering, Indian Institute of Technology Roorkee, Uttarakhand, India

Sandeep Gairola Centre of Excellence in Disaster Mitigation and Management, Indian Institute of Technology Roorkee, Roorkee, Uttarakhand, India

Vaibhav Gaur Academy of Scientific and Innovative Research (AcSIR), Ghaziabad, India;
CSIR-National Metallurgical Laboratory, Jamshedpur, India

Chiradeep Ghosh Research and Development Division, Tata Steel Limited, Jamshedpur, India

Megha Goyal Centre for Nanotechnology, Indian Institute of Technology Roorkee, Roorkee, India

Amit Kumar Gupta Department of Materials and Metallurgical Engineering, National Institute of Advanced Manufacturing Technology, Ranchi, India

R. K. Gupta ISRO-Vikram Sarabhai Space Centre, Trivandrum, India

Rajeev Gupta Materials Science Programme, IIT Kanpur, Kanpur, India

Shivani Gupta Mechanical and Industrial Engineering, Indian Institute of Technology Roorkee, Roorkee, India;
MIT World Peace University, Pune, India

Sajad Hamid Indian Institute of Technology, (IITR), Roorkee, Uttarakhand, India

B. Indranil Nuclear Fuel Complex, Hyderabad, India

Bhavya Jain Galgotias College of Engineering and Technology, Greater Noida, India

Jayant Jain Department of Materials Science and Engineering, IIT Delhi, New Delhi, India

P. Jeevanandam Department of Chemistry, Indian Institute of Technology Roorkee, Roorkee–247667, India

Arun Jena Department of Mechanical and Industrial Engineering, Indian Institute of Technology Roorkee, Uttarakhand, India

S. Anandh Jesuraj Centre for Nanoscience and Nanotechnology, Sathyabama Institute of Science and Technology, Chennai, India

Rohit Jha Department of Metallurgical Engineering, Indian Institute of Technology (Banaras Hindu University), Varanasi, India

Babu Rao Jinugu Department of Metallurgical Engineering, AU College of Engineering-Andhra University, Visakhapatnam, India

Komal Kapoor Nuclear Fuel Complex, Hyderabad, India

A. Karthikeyan School of Mechanical Engineering, Sathyabama Institute of Science and Technology, Chennai, India

Lokendra Kumar Katiyar Department of MME, Maulana Azad National Institute of Technology, Bhopal, MP, India

Pankaj Kaushik Department of Mechanical and Industrial Engineering, Indian Institute of Technology Roorkee, Uttarakhand, India

Mohd. Sarim Khan Department of MME, Maulana Azad National Institute of Technology, Bhopal, MP, India

A. M. Kamalan Kirubaharan Centre for Nanoscience and Nanotechnology, Sathyabama Institute of Science and Technology, Chennai, India;
Coating Department, Fun Glass - Centre for Functional and Surface Functionalised Glass, Alexander Dubcek University of Trencin, Trencin, Slovakia

Vikas Kukshal NIT Uttarakhand, Srinagar, Uttarakhand, India;
Department of Mechanical Engineering, National Institute of Technology Uttarakhand, Garhwal, India

Amit Kumar Department of Mechanical and Industrial Engineering, Indian Institute of Technology Roorkee, Roorkee, India

Binod Kumar Materials and Metallurgical Engineering Department, NIAMT, Hatia, Ranchi, India

Gaurav Kumar NIT Uttarakhand, Srinagar, Uttarakhand, India

Harish Kumar Mechanical Engineering Department, National Institute of Technology Delhi, New Delhi, India

Manoj Kumar Metal Extraction and Recycling Division, CSIR-National Metallurgical Laboratory(CSIR-NML), Jamshedpur, India

Mohit Kumar NIT Uttarakhand, Srinagar, Uttarakhand, India;
IIT Roorkee, Roorkee, Uttarakhand, India

Mukund Kumar NIT Jamshedpur, Jamshedpur, Jharkhand, India

Nitin Kumar Indian Institute of Technology Roorkee, Roorkee, India

R. Sunil Kumar Department of Metallurgical & Materials Engineering, National Institute of Technology Andhra Pradesh, Tadepalligudem, India

Sumit Kumar Department of Metallurgical and Materials Engineering, Indian Institute of Technology Roorkee, Roorkee, Uttarakhand, India

V. Anil Kumar ISRO-Vikram Sarabhai Space Centre, Trivandrum, India

Debrupa Lahiri Indian Institute of Technology Roorkee, Roorkee, India

Gaurav Mahendru Applied Materials India Pvt. Ltd., Bangalore, India

Gopi Kishor Mandal Academy of Scientific and Innovative Research (AcSIR), Ghaziabad, India;
CSIR-National Metallurgical Laboratory (CSIR-NML), Jamshedpur, India

Tapas Kumar Mandal Centre for Nanotechnology, Indian Institute of Technology Roorkee, Roorkee, India;
Department of Chemistry, Indian Institute of Technology Roorkee, Roorkee, India

Tara Chand Kumar Maurya Indian Institute of Technology, Roorkee, Uttarakhand, India

Pavan Meena Department of Materials and Metallurgical Engineering, Maulana Azad National Institute of Technology, Bhopal, MP, India

Arunabh Meshram Department of Materials Science & Engineering, IIT Kanpur, Kanpur, India

Joy Prakash Misra Indian Institute of Technology BHU, Varanasi, India

Girija Moona CSIR–National Physical Laboratory, New Delhi, India

Pravin Muneshwar Materials Processing Division, Vikram Sarabhai Space Centre, Trivandrum, India

Leeladhar Nagdeve Mechanical Engineering Department, National Institute of Technology Delhi, New Delhi, India

Monika Narwal Department of Chemistry, Indian Institute of Technology Roorkee, Roorkee–247667, India

S. K. Nath Indian Institute of Technology Roorkee, Roorkee, India

Sumeer K. Nath Department of Metallurgical and Materials Engineering, Indian Institute of Technology Roorkee, Roorkee, Uttarakhand, India

Subha Sanket Panda Department of Materials Science and Engineering, IIT Kanpur, Kanpur, India

Bhanu Pant Kalyani Centre for Technological Innovation, Pune, India

Meena Pant Mechanical Engineering Department, National Institute of Technology Delhi, New Delhi, India

D. V. T. Pardhasaradhi Department of Metallurgical & Materials Engineering, National Institute of Technology Andhra Pradesh, Tadepalligudem, India

Krishnakant Phand Indian Institute of Technology Roorkee, Roorkee, India

K. Prabhakaran Department of Chemistry, Indian Institute of Space Science and Technology, Thiruvananthapuram, India

Guru Prakash Indian Institute of Technology Roorkee, Roorkee, India;
Metallurgical Engineering Department, O P Jindal University, Raigarh, India

Ujjwal Prakash Indian Institute of Technology, (IITR), Roorkee, Uttarakhand, India

K. Anusha Raj Department of Metallurgical and Materials Engineering, Rajiv Gandhi University of Knowledge Technologies, RK Valley Campus, Kadapa, India

Maran Rajakumaran Department of Mechanical and Industrial Engineering, Indian Institute of Technology Roorkee, Roorkee, Uttarakhand, India

M. Rajasekaramoorthy School of Mechanical Engineering, Sathyabama Institute of Science and Technology, Chennai, India;
Centre for Nanoscience and Nanotechnology, Sathyabama Institute of Science and Technology, Chennai, India

V. Rajinikanth Materials Engineering Division, CSIR-National Metallurgical Laboratory, Jamshedpur, India

M. V. Ramana Nuclear Fuel Complex, Hyderabad, India

G. Ramesh Department of Metallurgical and Materials Engineering, Rajiv Gandhi University of Knowledge Technologies, RK Valley Campus, Kadapa, India

Sandeep Sahni Department of Materials Science and Engineering, IIT Kanpur, Kanpur, India

Lipak Kumar Sahoo Department of Metallurgical and Materials Engineering, Indian Institute of Technology Madras, Chennai, India

Vikash Kumar Sahu Academy of Scientific and Innovative Research (AcSIR), Ghaziabad, India;
CSIR-National Metallurgical Laboratory (CSIR-NML), Jamshedpur, India

M. Sambasiva Rao Foundry Forge Technology Department, National Institute of Advanced Manufacturing Technology (Formerly-NIFFT), Ranchi, India

Sabita Sarkar Department of Metallurgical and Materials Engineering, Indian Institute of Technology Madras, Chennai, India

C. Sasikumar Department of Materials and Metallurgical Engineering MANIT Bhopal, Bhopal, MP, India

T. Savanth Department of Metallurgical & Materials Engineering, National Institute of Technology Andhra Pradesh, Tadepalligudem, India

Apurbba Kumar Sharma Mechanical and Industrial Engineering, Indian Institute of Technology Roorkee, Roorkee, India

Ganesh Kumar Sharma Department of Mechanical Engineering, National Institute of Technology Uttarakhand, Garhwal, India;
Department of Mechanical Engineering, Moradabad Institute of Technology, Moradabad, India

Gaurav Sharma Department of Mechanical Engineering, Birla Institute of Technology and Science (BITS) Pilani, Hyderabad Campus, Secunderabad, Telengana, India

Hitesh Sharma National Institute of Technology, Srinagar, Uttarakhand, India

Rahul Sharma Department of Materials and Metallurgical Engineering MANIT Bhopal, Bhopal, India

Hafeez Shekh Department of Metallurgical and Materials Engineering, Indian Institute of Technology Roorkee, Roorkee, Uttarakhand, India

Inderdeep Singh Department of Mechanical and Industrial Engineering, Indian Institute of Technology Roorkee, Roorkee, India;
Centre of Excellence in Disaster Mitigation and Management, Indian Institute of Technology Roorkee, Roorkee, Uttarakhand, India

Kamalesh K. Singh Department of Metallurgical Engineering, Indian Institute of Technology (Banaras Hindu University), Varanasi, India

Pravendra Pratap Singh Metallurgical and Materials Engineering, Indian Institute of Technology Roorkee, Roorkee, India;
Department of Chemistry, Indian Institute of Space Science and Technology, Thiruvananthapuram, India

Rajdev Singh Department of Mechanical and Industrial Engineering, Indian Institute of Technology Roorkee, Uttarakhand, India

Randhir Singh Department of Metallurgical Engineering, Indian Institute of Technology (Banaras Hindu University), Varanasi, India

Sudhanshu Shekhar Singh Department of Materials Science and Engineering, IIT Kanpur, Kanpur, India

Shishir Sinha Centre of Excellence in Disaster Mitigation and Management, Indian Institute of Technology Roorkee, Roorkee, Uttarakhand, India;
Department of Chemical Engineering, Indian Institute of Technology Roorkee, Roorkee, Uttarakhand, India

Dinesh Srivastava Nuclear Fuel Complex, Hyderabad, India

V. C. Srivastava Materials Engineering Division, CSIR-National Metallurgical Laboratory, Jamshedpur, India

M. Subhakar Department of Materials Science and Engineering, IIT Delhi, New Delhi, India

S. Subrahmanyam Department of Metallurgical & Materials Engineering, National Institute of Technology Andhra Pradesh, Tadepalligudem, India

G. Sugilal Homi Bhabha National Institute, Mumbai, India;
Bhabha Atomic Research Center, Mumbai, India

Kurra Suresh Department of Mechanical Engineering, Birla Institute of Technology and Science (BITS) Pilani, Hyderabad Campus, Secunderabad, Telengana, India

Chandra Sekhar Tiwary Department of Mining Engineering, Indian Institute of Technology, Kharagpur, West Bengal, India

Snehashish Tripathy Academy of Scientific and Innovative Research (AcSIR), Ghaziabad, India;
CSIR-National Metallurgical Laboratory (CSIR-NML), Jamshedpur, India

Harish Trivedi Materials Science Programme, IIT Kanpur, Kanpur, India

Anand Udare College of Engineering, Pune, India

Anu Verma School of Environmental Science and Engineering, Indian Institute of Technology, Kharagpur, West Bengal, India

Amol Vuppuluri Department of Mechanical Engineering, Birla Institute of Technology and Science (BITS) Pilani, Hyderabad Campus, Secunderabad, Telengana, India

Praveen Wilson Department of Chemistry, Indian Institute of Space Science and Technology, Thiruvananthapuram, India

Ajaykumar Udayraj Yadav Materials and Metallurgical Engineering Department, NIAMT, Hatia, Ranchi, India

Akeshwar Singh Yadav Jindal Steel and Power Ltd., Raigarh, CG, India;
Indian Institute of Technology Roorkee, Roorkee, India

Part I
Advanced Materials Processing

Chapter 1
Surface Modification of NiO Nanoparticles Using Stearic Acid and Their Application as Adsorbent

Monika Narwal and P. Jeevanandam

Abstract NiO is a p-type semiconductor which shows interesting catalytic, electrical and magnetic properties due to finite size effects. NiO nanoparticles are used in various applications such as adsorption, supercapacitors, Li-ion batteries and catalysis. NiO nanoparticles have some limitations as an adsorbent. To overcome these limitations, surface of NiO nanoparticles has been modified with different surface modifying agents such as stearic acid, silane coupling agents, dodecanoic acid, etc. The surface modifiers chemically alter the surface properties of NiO nanoparticles without affecting their bulk properties. The surface-modified NiO nanoparticles can be employed in various applications such as adsorption, heavy metal ions removal and oil–water separation. In the current study, NiO nanoparticles have been synthesized using sol–gel method. The synthesized NiO nanoparticles were characterized using various analytical techniques such as X-ray diffraction, infrared spectroscopy and electron microscopy. Then, a simple and cost-effective strategy was used to modify the surface of NiO nanoparticles using stearic acid (SA). Varying amounts of stearic acid were used to modify the surface of NiO nanoparticles. The surface modification of NiO nanoparticles with stearic acid was confirmed by different analytical techniques such as FT-IR, TGA and CHO analyses. FT-IR analysis shows characteristic IR bands due to stearic acid confirming the presence of stearic acid on the surface of NiO nanoparticles. TGA and CHO analyses also confirm the presence of stearic acid on the surface of NiO nanoparticles. The stearic acid-modified NiO nanoparticles were tested as adsorbent and they perform as better adsorbent for the removal of crystal violet (CV) from an aqueous solution compared to pure NiO nanoparticles.

Keywords Sol–gel method · Surface modification · Adsorption

M. Narwal · P. Jeevanandam (✉)
Department of Chemistry, Indian Institute of Technology Roorkee, Roorkee–247667, India
e-mail: jeevafcy@iitr.ac.in

A. Tewari et al. (eds.), *Proceedings of the 3rd International Conference on Advances in Materials Processing: Challenges and Opportunities*, Springer Proceedings in Physics 293, https://doi.org/10.1007/978-981-99-1971-0_1

1.1 Introduction

Nickel oxide is a material with good electrochemical stability, biocompatibility, catalytic activity, interesting magnetic properties and low production cost [1]. NiO is commonly employed in various applications such as dye-sensitized photoelectrodes, electrochromic sheets, anodes, fuel cells, adsorption, catalysis and smart windows [2]. NiO nanoparticles often do not have suitable surface properties for specific applications such as adsorption and photocatalytic degradation [3]. Another issue is the agglomeration of NiO nanoparticles due to their high surface-to-volume ratio and high surface energy. This problem can be solved by modifying the surface of NiO nanoparticles. The organic compounds that are frequently used to modify the surface of metal oxide nanoparticles include thiols, carboxylic acids, phosphonates, silanes and amines [4]. Surface modification of NiO nanoparticles with organic acids minimizes the surface energy and increases their stability. Stearic acid has been used to reduce surface energy and to convert hydrophilic nanoparticles to hydrophobic ones [5].

In the present study, the surface of NiO nanoparticles was modified with stearic acid at room temperature using a low-cost and simple approach. The surface-modified NiO nanoparticles were tested as adsorbent for the adsorption of crystal violet from an aqueous solution.

1.2 Experimental Details

The chemicals used were nickel acetate (Aldrich), toluene (SRL), stearic acid (Himedia) and ethanol (Changshu Hongsheng Fine Chemical Co.Ltd.). All the chemicals were used as received (without further purification). NiO nanoparticles were synthesized using sol–gel method [6]. In the synthesis, 10 mmol of nickel acetate was dissolved in 80 mL of ethanol followed by addition of 1 mL of water with constant magnetic stirring at RT for 3 h. Then, a mixture of toluene (50 mL) and NH_4OH solution (4 mL) was added to the resultant solution with constant stirring followed by the addition of water (2 mL) after 15 minutes. The contents were stirred at RT for 23 h to obtain a sol. A gel was obtained by heating the sol at 60 °C. The obtained gel was dried in an oven overnight. The obtained product was calcined at 500 °C in a muffle furnace to get NiO nanoparticles. The synthesized NiO nanoparticles were modified using stearic acid (SA). For this, different amounts of stearic acid (15–45 mg) and 50 mg of NiO nanoparticles were taken in 20 mL of ethanol with constant stirring for 5 minutes. This is followed by centrifugation and washing with ethanol. The SA-modified NiO nanoparticles prepared using different amounts of SA, i.e. 15 mg, 30 mg and 45 mg are designated as NiO(SA1), NiO(SA2) and NiO(SA3), respectively.

1.3 Results and Discussion

XRD patterns of as-prepared and calcined (500 °C) NiO samples are shown in Fig. 1.1a.

The XRD pattern of as-prepared NiO matches with that of α-Ni-$(OH)_2$ (JCPDS file #. 38−0715) and the pattern of calcined NiO matches with that of NiO (JCPDS file #. 47−1049) [7]. The estimated crystallite size of NiO nanoparticles is 13.1 nm. TEM images of stearic acid-modified and unmodified NiO nanoparticles (Fig. 1.1b) show that there is no change in the morphology of NiO nanoparticles after modification with stearic acid. The average size of unmodified and modified NiO nanoparticles is 14.4 ± 2.7 nm and 13.2 ± 1.6 nm, respectively.

The surface modification of NiO nanoparticles with stearic acid was confirmed by FT-IR, TG and CHO analyses. Figure 1.2a shows the FT-IR spectra of pure stearic acid, pure NiO and stearic acid-modified NiO nanoparticles. In pure stearic acid, the IR bands at 2928 cm^{-1} and 2852 cm^{-1}, and 1705 cm^{-1} correspond to sym. and asym. stretching of $-CH_2$ groups, and stretching of free COOH, respectively. In pure NiO nanoparticles, the IR band around 480 cm^{-1} is due to stretching of Ni–O bond. The

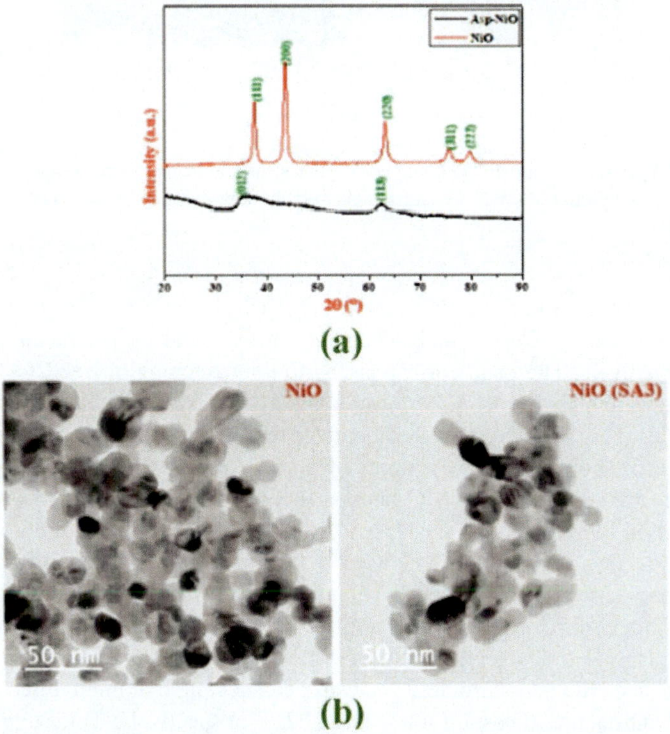

Fig. 1.1 **a** XRD patterns of as-prepared and calcined NiO nanoparticles and **b** TEM images of unmodified and stearic acid-modified NiO nanoparticles

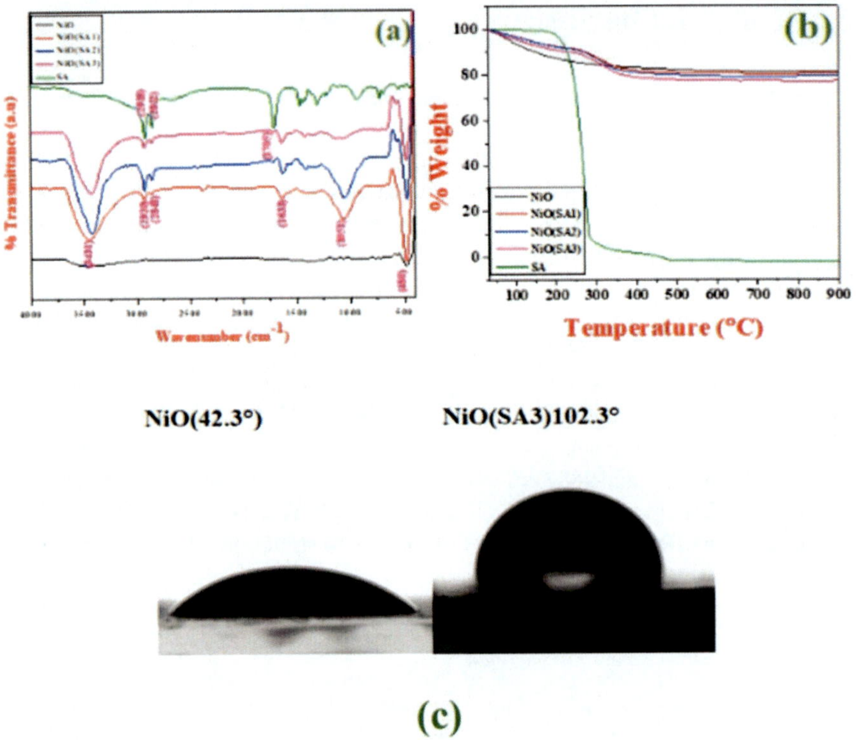

Fig. 1.2 **a** FT-IR spectra and **b** TGA curves of pure stearic acid, unmodified and stearic acid-modified NiO nanoparticles. **c** Water contact angle measurements of unmodified and modified NiO nanoparticles

FT-IR spectra of SA-modified NiO samples (NiO(SA1), NiO(SA2) and NiO(SA3)) show bands around 2920 cm^{-1} and 2848 cm^{-1} due to sym. and asym. stretching of CH_2 group, respectively. The band corresponding to the stretching of free COOH shifts to lower wavenumber from 1705 cm^{-1} to about 1633 cm^{-1} in the SA-modified NiO nanoparticles. This suggests that SA is bound to the surface of NiO nanoparticles [8].

Figure 1.2b shows the TGA patterns of pure stearic acid, pure NiO nanoparticles and stearic acid-modified NiO nanoparticles. The TGA results show that the decomposition temperature of stearic acid is increased in NA(SA) as compared to pure SA [9]. The CHO analysis showed that after stearic acid modification, the percentage of carbon increases in SA-modified NiO nanoparticles (from 0.03 to 2.87 %). This confirms the modification of NiO nanoparticles with stearic acid. Figure 1.2c shows the water contact angle measurements of stearic acid-modified and unmodified NiO nanoparticles. The water contact angle of unmodified and SA-modified NiO nanoparticles are 42.3° and 102.3°, respectively. This suggests that the surface of unmodified NiO is hydrophilic, while the surface of SA-modified NiO nanoparticles is hydrophobic [10].

Figure 1.3a shows UV–Visible spectral results on the adsorption of crystal violet (CV), a toxic dye, from solution. For this, 10 mg of adsorbent was dispersed in 10 mL of aqueous crystal violet solution (2×10^{-5} M). The mixture was sonicated and kept at RT for 30 min. The mixture was centrifuged and the concentration of crystal violet in the supernatant solution was analysed using a UV–Vis spectrophotometer. The UV–Vis spectral results show that all the SA-modified NiO nanoparticles show better adsorption of crystal violet from the solution as compared to the unmodified NiO nanoparticles. Figure 1.3b shows the kinetics of adsorption of CV using SA-modified and unmodified NiO nanoparticles. The kinetic results show that stearic acid-modified NiO nanoparticles show complete adsorption of CV within 5 min. Thus, the stearic acid modified NiO nanoparticles act as good adsorbent for the removal of crystal violet. The adsorption of crystal violet is due to hydrophobic interaction between the SA-modified NiO nanoparticles and CV as shown in Figure 1.3c [11].

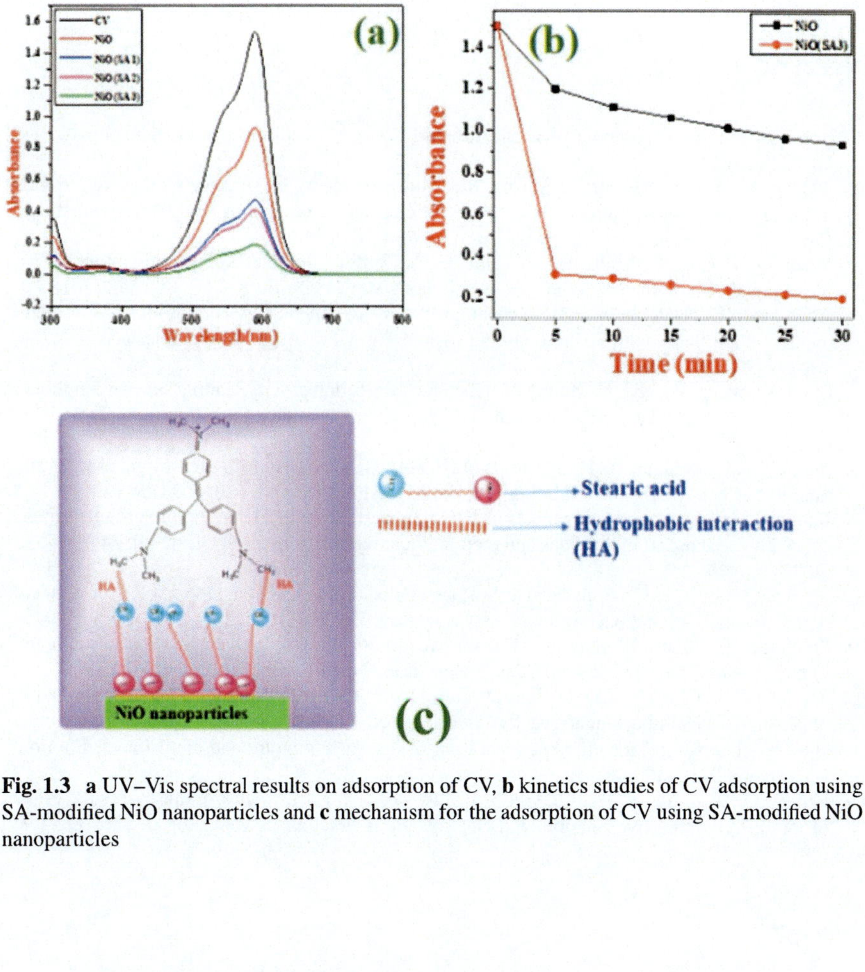

Fig. 1.3 **a** UV–Vis spectral results on adsorption of CV, **b** kinetics studies of CV adsorption using SA-modified NiO nanoparticles and **c** mechanism for the adsorption of CV using SA-modified NiO nanoparticles

1.4 Conclusions

In the current study, NiO nanoparticles were synthesized using sol–gel method and the nanoparticles were surface modified with stearic acid in 5 minutes at RT. The surface modification of NiO nanoparticles with stearic acid was confirmed by FT-IR analysis, TGA and CHO analyses. The stearic acid-modified NiO nanoparticles show higher water contact angle than the unmodified NiO nanoparticles suggesting their hydrophobic nature. The surface-modified NiO nanoparticles perform as better adsorbent for CV removal from solution compared to the unmodified NiO nanoparticles.

Acknowledgments Monika Narwal gratefully acknowledges the Council of Scientific and Industrial Research (CSIR), Government of India, for providing fellowship (JRF/SRF). The authors are grateful to the Institute Instrumentation Centre, IIT Roorkee, for providing various facilities.

References

1. Bonomo, M.: Synthesis and characterization of NiO nanostructure: a review. J. Nanopart. Res. **20**, 222 (2018)
2. Ahmad, W., Chandra Bhatt, S., Verma, M., Kumar, V., Kim, H.: A review on current trends in the green synthesis of nickel oxide nanoparticles, characterizations, and their applications. Environ. Nanotechnol. Monit. Manag. **18**, 100674 (2022)
3. Yazdi, G., Ivanic, M., Mohamed, A., Uheida, A.: Surface modified composite nanofibers for the removal of indigo carmine dye from polluted water. RSC Adv. **8**, 24588–24598 (2018)
4. Ahangaran, F., Navarchian, A.H.: Recent advances in chemical surface modification of metal oxide nanoparticles with silane coupling agents: a review. Adv. Colloid Interface Sci. **286**, 102298 (2020)
5. Chen, X., Wang, M., Xin, Y., Huang, Y.: One-step fabrication of self-cleaning superhydrophobic surfaces: a combined experimental and molecular dynamics study. Surf. Interfaces **31**, 102022 (2022)
6. Yadav, S.K., Jeevanandam, P.: Synthesis of NiO-Al_2O_3 nanocomposites by sol-gel process and their use as catalyst for the oxidation of styrene. J. Alloys Compd. **610**, 567–574 (2014)
7. Rana, P., Jeevanandam, P.: Synthesis of NiO nanoparticles via calcination of surfactant intercalated layered nickel hydroxides and their application as adsorbent. J. Clust. Sci. **34**, 517–533 (2022)
8. Singh, S., Barick, K.C., Bahadur, D.: Surface engineered magnetic nanoparticles for removal of toxic metal ions and bacterial pathogens. J. Hazard. Mater. **192**, 1539–1547 (2011)
9. Soleimani, E., Taheri, R.: Synthesis and surface modification of CuO nanoparticles: evaluation of dispersion and lipophilic properties. Nano-Struct. Nano-Objects **10**, 167–175 (2017)
10. Xu, X., Li, X., Chen, Y., Liu, G., Gao, J., Liu, J., Liu, Z., Yue, S., Zhang, L.: Facile fabrication of 3D hierarchical micro-nanostructure fluorine-free superhydrophobic materials by a simple and low-cost method for efficient separation of oil-water mixture and emulsion. J. Environ. Chem. Eng. **9**, 106400 (2021)
11. Hu, Y., Guo, T., Ye, X., Li, Q., Guo, M., Liu, H., Wu, Z.: Dye adsorption by resins: effect of ionic strength on hydrophobic and electrostatic interactions. Chem. Eng. J. **228**, 392–397 (2013)

Chapter 2
Microstructure and Mechanical Properties of Dual Two-Phase (B2+DO_{24}) $Ti_{45}Fe_5Ni_{50}$ Intermetallic Alloy

Subha Sanket Panda, Sandeep Sahni, M. Subhakar, Jayant Jain, and Sudhanshu Shekhar Singh

Abstract Mechanical properties of a dual two-phase $Ti_{45}Fe_5Ni_{50}$ intermetallic alloy, comprising of matrix with B2 crystal structure and second-phase (volume fraction:~10 %) with DO_{24} crystal structure, have been investigated using tensile test, nanoindentation, and molecular dynamics (MD) simulation. The alloy exhibited an ultimate tensile strength (UTS) of 313 MPa and strain less than ~6 %. Both phases exhibited the same hardness; however, DO_{24} phase was found to possess a higher modulus than the B2 matrix. MD simulation suggested that the dislocation density in the DO_{24} phase accommodated the deformation and delayed the damage process.

Keywords Tensile testing · Nanoindentation · Molecular dynamic simulation

2.1 Introduction

Multi-phase intermetallic alloys have been considered as novel materials, which possess sufficient strength and stability under adverse operating conditions in addition to maintaining adequate amount of toughness at ambient temperatures, and thus ensuring structural integrity [1]. Hypereutectoid alloy compositions have been typically used as dual two-phase intermetallic alloys in many binary systems, e.g., NiAl, TiAl, and MoSi. Dual two-phase intermetallic alloys exhibit crystallographic coherence between the constituent phases and are very much stable at elevated temperatures [2]. Kawahara et al. [3] studied the variation of hardness in dual two-phase intermetallic alloy comprising of Ni_3Al and Ni_3V phases as a function of Al content and observed that the increase in the volume fraction of Ni_3Al resulted in a decrease in the hardness. In addition, Ni_3Al was also found to be the primary factor influencing the mechanical properties. Appel and Christoph [4] studied the coherency

S. S. Panda · S. Sahni · S. S. Singh (✉)
Department of Materials Science and Engineering, IIT Kanpur, Kanpur, India
e-mail: sudhanss@iitk.ac.in

M. Subhakar · J. Jain
Department of Materials Science and Engineering, IIT Delhi, New Delhi, India

A. Tewari et al. (eds.), *Proceedings of the 3rd International Conference on Advances in Materials Processing: Challenges and Opportunities*, Springer Proceedings in Physics 293, https://doi.org/10.1007/978-981-99-1971-0_2

stress and interface-related deformation in intermetallic alloy comprising of alternate lamella of TiAl and Ti_3Al and found that the stress existing at the interface due to lattice mismatch gives rise to the internal stress field which boasts dislocation generation resulting in ductility and beneficial damage tolerance. Ilyas and Kabir [5] modeled the anisotropy in creep behavior of two-phase (TiAl and Ti_3Al) intermetallic alloy using crystal plasticity simulation and observed that, depending upon lamellar orientations, locally deformation modes get active which affects the creep properties globally. In most cases, studies have been carried out on the binary system. To further enhance their performance, transition metals are being considered as possible ternary additions which ensure microstructural stability as well as both solid solution and precipitation strengthening. Due to their attractive properties and significant structural applications, ternary additions have been made in titanium aluminide and nickel aluminide-based dual two-phase intermetallic alloys in the past [6]. It is not uncommon that the mechanical properties of such multi-phase intermetallic alloys depend upon the fraction of the constituent phases, properties of the individual phase, and crystallographic coherence/semi-coherence across the interface. There exist several binary-ternary systems with ternary-quaternary substitutions containing two/multi-intermetallic phases having different crystal structures. Thus, it is imperative to study the mechanical properties of such systems and identify the underlying deformation mechanisms. In this work, the authors have studied the microstructure and mechanical properties of dual two-phase intermetallic alloy with a nominal composition of $Ti_{45}Fe_5Ni_{50}$.

2.2 Materials and Methods

Ti–Fe–Ni intermetallic alloy with nominal composition $Ti_{45}Fe_5Ni_{50}$ in atomic percent (at.%) was fabricated via conventional arc melting technique in an argon atmosphere (10^{-7} Torr) using a non-consumable tungsten electrode. Subsequently, the cast materials were homogenized at a temperature of 1273 K for 50 h followed by furnace cooling. Microstructural characterization was carried out using scanning electron microscopy (SEM) equipped with energy dispersive spectroscopy (EDS). Phase identification was carried out using X-ray diffraction (XRD). For mechanical properties evaluation, tensile test and nanoindentation were carried out. The tensile test was conducted at room temperature at a strain rate of 3×10^{-4} s^{-1}. The nanoindentation test was performed on both matrix and second phase using a Berkovich Indenter at loads of 100 mN and 30 mN, respectively. To understand the dislocation mechanism during deformation, molecular dynamics (MD) simulation was carried out using LAMMPS on two-phase nano poly-crystalline (NPC) intermetallic alloys. Dislocation evolution was studied using the dislocation extraction algorithm (DXA). The atomic configurations and their evolutions were analyzed using common neighbor analysis (CNA) and visualized using open visualization tool (OVITO) software.

2.3 Results and Discussion

Figure 2.1a shows back-scattered (BSE) image of the homogenized intermetallic alloy. It can be observed that the microstructure consists of two phases, i.e., plate or needle-like and globular-shaped relatively large (Ni, Fe)$_3$Ti (Crystal structure: DO$_{24}$) phase is dispersed in the randomly oriented B2 matrix. The XRD pattern shown in Fig. 2.1b also confirms the presence of primarily two phases along with (Ni, Fe)$_3$Ti$_2$. The volume fraction of (Ni, Fe)$_3$Ti was calculated using ImageJ software and was found to be ~10 %. Figure 2.1c shows the engineering stress–strain curve of the intermetallic under tension. The ultimate tensile strength and strain to failure were found to be 313 $\pm$ 50 MPa and 5.1 $\pm$ 0.9 %, respectively. It is to be noted that the strain was calculated without an extensometer; therefore, the strain is slightly overestimated. Figure 2.1d shows the room temperature hardness values of each phase obtained from nanoindentation. The hardness and modulus of the B2 matrix were found to be 5.65 $\pm$ 0.27 GPa and 97.5 $\pm$ 2.6 GPa, respectively, whereas the hardness and modulus of (Ni, Fe)$_3$Ti phase were measured to be 5.69 $\pm$ 0.45 GPa and 142.0 $\pm$ 17.0 GPa, respectively. It can be observed that there is no significant difference in the hardness of the matrix and second phase. The observed higher modulus for the second phase can be attributed to the presence of higher Ni and low Ti content. The hardness of the (Ni, Fe)$_3$Ti phase can be attributed to the topologically close-packed crystal structure having a double hexagonal stacking sequence of ABACABACABAC and more atoms per unit cell. Similarly, a high hardness value for the matrix could be attributed to the presence of higher Ti content despite a relatively simple and higher symmetry crystal structure.

In order to understand the deformation behavior of the dual-phase intermetallic system, atomistic simulations were carried out. It is not appropriate to explain the bulk tensile behavior of the material from atomistic scale; however, dislocation behavior during deformation can be effectively used to explain the deformation characteristics irrespective of the length scale. Figure 2.2(a1–a3) shows the simulation box for each phase B2 and DO$_{24}$ and two-phase (B2 + DO$_{24}$) considered for the calculation. Figure 2.3(a1–a3) and (b1–b3) show the evolution of dislocations during tensile deformation for DO$_{24}$ and B2 phases, respectively. It can be observed that, at the initial stages of deformation, the dislocation density in the (Ni, Fe)$_3$Ti phase is higher than the B2 matrix. With subsequent deformation, the dislocation density in each phase increased; however, the matrix exhibits lower dislocation density as compared to the (Ni, Fe)$_3$Ti phase even at higher extent of deformation. Thus, it can be said that despite having a low volume fraction of the second phase, it contributed significantly towards plastic strain, as explained using deformation mechanisms in the next section. Further dislocation characterization was carried out using DXA, where basal {0001} slip was observed to be the dominant slip for the (Ni, Fe)$_3$Ti phase with majority of dislocations were having Burgers vector 1/3<1$\bar{2}$10> and 1/3<1$\bar{1}$00>. The rationale behind the activation of basal slip is attributed to its crystal structure, which is intermediate between FCC and HCP [7]. Therefore, it might be assumed that the super intrinsic stacking fault energy (SISF) is not very high for the (Ni, Fe)$_3$Ti phase.

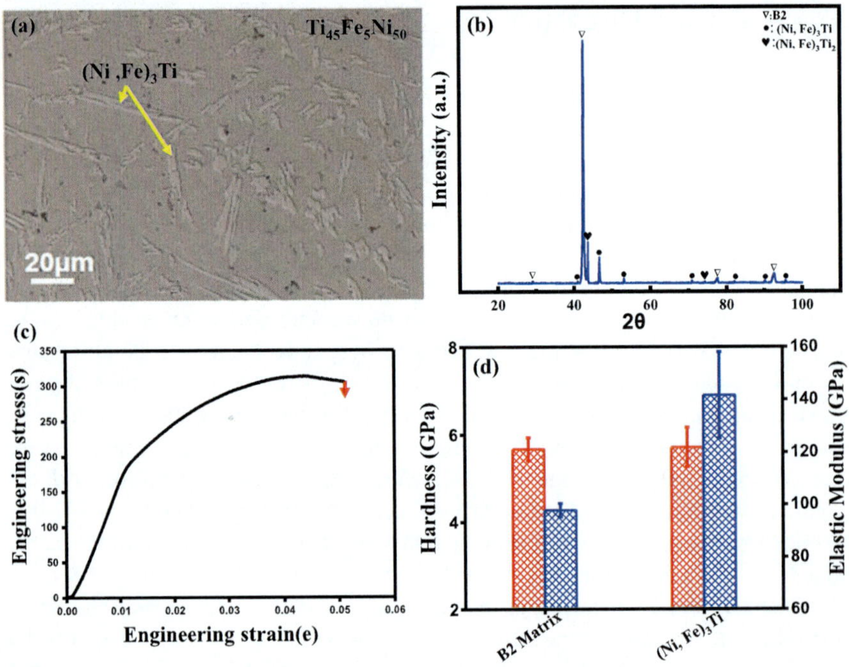

Fig. 2.1 **a** BSE image of the homogenized alloy, **b** X-ray diffraction spectrum, **c** stress versus strain curve under tensile loading, and **d** hardness and elastic modulus for both phases

It should be noted that no pyramidal slip was observed which could be ascribed to the complicated arrangement of atoms in the pyramidal plane. For B2 matrix, the predominant slip system was found to be $\{1\bar{1}0\}1/2 <111>$. As mentioned before, with subsequent deformation, the increase in the dislocation density in the B2 matrix was observed to be lower as compared to the $(Ni, Fe)_3Ti$ phase. This indicates that B2 matrix is not able to accommodate significant strain, which might be ascribed to its high antiphase boundary (APB) energy [8]. In the case of $(Ni, Fe)_3Ti$ phase, SISF energy is not very high, and therefore it is able to accommodate more deformation.

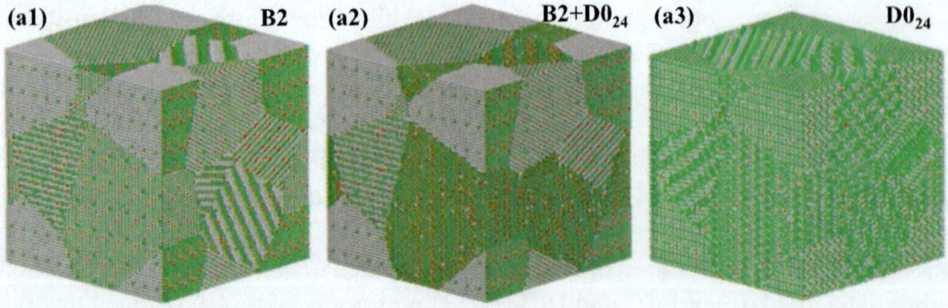

Fig. 2.2 Simulation box individual phase (a1: B2, a3: DO_{24}) as well as two phase (a2: B2+ DO_{24})

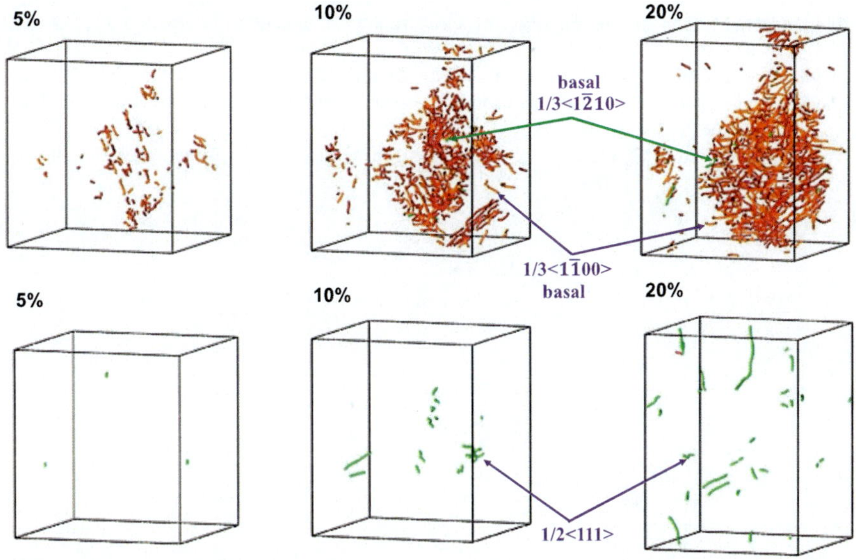

Fig. 2.3 Evolution of dislocation during tensile deformation using MD simulation in (a1–a3) DO_{24} and (b1–b3) B2 crystal structure

2.4 Conclusions

The mechanical behavior of dual two-phase (B2+ DO_{24}) $Ti_{45}Fe_5Ni_{50}$ intermetallic alloy has been studied experimentally as well as using atomistic simulation. The plastic deformation was observed to be predominantly dominated by the DO_{24} phase despite its low volume fraction. The matrix with B2 crystal structure, owing to its highly stable APB energy, did not deform much, which eventually resulted in failure.

References

1. Ohira, K., Kaneno, Y., Takasugi, T.: Microstructure, mechanical property, and oxidation property in Ni_3Si-Ni_3Ti-Ni_3Nb multiphase intermetallic alloys. Mater. Sci. Eng. A. **399**, 332–343 (2005)
2. Soga, W., Kaneno, Y., Yoshida, M., Takasugi, T.: Microstructure and mechanical property in dual two-phase intermetallic alloys composed of geometrically close packed Ni_3X (X: Al and V) containing Nb. Mater. Sci. Eng. A. **473**, 180–188 (2008)
3. Kawahara, K., Kaneno, Y., Kakitsuji, A and Takasugi. T.: Microstructural factors affecting hardness property of dual two-phase intermetallic alloys based on Ni_3Al-Ni_3V pseudo-binary alloy system. Intermet. **17**, 938-944 (2009)
4. Appel, F., Chrishtoph, A.: Coherency stresses and interface related deformation phenomena in two-phase titanium aluminides. Intermet. **7**, 113–1182 (1999)
5. Umer Ilyes, M., Rizviul Kabir, M.: Creep behavior of two phase lamellar TiAl: crystal plasticity modelling and analysis. Intermet. **132**, 107129 (2021)

6. Sauthoff, G.: Multiphase intermetallic alloys for structural applications. Intermet. **8**, 1101–1109 (2000)
7. Hagihara, K., Nakano, T., Umakoshi, Y.: Plastic deformation behavior of Ni_3X (X = Nb, Ti and Sn) type HCP based Intermetallics with the geometrically close packed structure. Mat. Res. Soc. Symp. Proc. **753**
8. Yoo, M.H., Takasugi, T., Hanada, H., Izumi, O.: Slip modes in B2 type intermetallic alloys. Mater. Trans., JIM. **6**(31), 435–442 (1990)

Chapter 3
Finite Element Analysis of Melting of Bulk Metals Using Microwave Energy at 2.45 GHz

Gaurav Kumar, Mohit Kumar, Vikas Kukshal, and Mukund Kumar

Abstract Microwave material processing is a rapid and cost-effective method with significant advantages over conventional methods of material processing. The present study focuses on the finite element analysis of melting of bulk metals such as aluminum alloy (Al-7039), magnesium alloy (Mg-AZ31B), and lead using microwave energy at a maximum input power of 1400 W. The simulation study was carried out using COMSOL Multiphysics. A silicon carbide susceptor was used for initial microwave coupling and heating the bulk metal via the conventional mode of heat transfer. Once the bulk metal reaches a critical temperature, the bulk metal gets heated using both conventional and microwave modes of heating. The finite element analysis has been used to describe the time at which these bulk metals start melting. The effect of different input power and casting volume on electric field distribution and time–temperature profile of metals has also been discussed.

Keywords Microwave material processing · COMSOL multiphysics · Melting

3.1 Introduction

The advanced manufacturing process focuses on improving industry performance through cost optimization, improved product quality, low energy consumption, less time consumption, product flexibility, and less environmental hazard [1, 2]. Microwave heating is the conversion of electromagnetic energy to thermal energy, which results in uniform, volumetric, and rapid heating throughout the material as

G. Kumar (✉) · M. Kumar · V. Kukshal
NIT Uttarakhand, Srinagar, Uttarakhand, India
e-mail: grv.kmr@nituk.ac.in

M. Kumar
IIT Roorkee, Roorkee, Uttarakhand, India

M. Kumar
NIT Jamshedpur, Jamshedpur, Jharkhand, India

A. Tewari et al. (eds.), *Proceedings of the 3rd International Conference on Advances in Materials Processing: Challenges and Opportunities*, Springer Proceedings in Physics 293, https://doi.org/10.1007/978-981-99-1971-0_3

well as a reduced thermal gradient [3, 4]. Because of its numerous advantages, microwave is used in a variety of processes such as sintering, joining, drilling, cladding, melting, and casting, among others, with or without the use of Microwave Hybrid Heating (MHH) [5–9]. MHH is a rapid heating process in which a susceptor is used to heat the bulk metals or microwave-reflecting materials initially, and then the material is heated by a combined action of conventional heating and microwave heating [10]. Singh et al. reported the fabrication of metal–ceramic composite casting of nickel (matrix) and Silicon carbide (reinforcement) powder through MHH at an input power of 900W [11]. Mishra and Sharma reported the in situ microwave casting of aluminum alloy-7039 at an input power of 1400W. The cast sample was having dense microstructure with <2% porosity [12]. Several authors have reported heating of copper, aluminum alloy-7039, tin, etc., and compared the experimental temperature profile with simulated temperature profile [13–15]. The temperature profile of microwave heating revealed that microwave melting was more efficient and took less time than conventional melting [15]. The objective of the present study is to study the effect of the power on the electric field inside the microwave oven and the melting time–temperature profile of lead and magnesium.

3.2 Material and Modeling

Figure 3.1a shows a 3D model of the microwave melting setup. The lower and upper molds, sprue, pouring basin, and susceptor were made of graphite, alumina, and Silicon Carbide (SiC), respectively. Table 3.1 shows the material properties. Free tetrahedral element was used for meshing over the whole domain as shown in Fig 3.1b. The number of domain elements, boundary elements, and edge elements was 128384, 12769, and 1215, respectively.

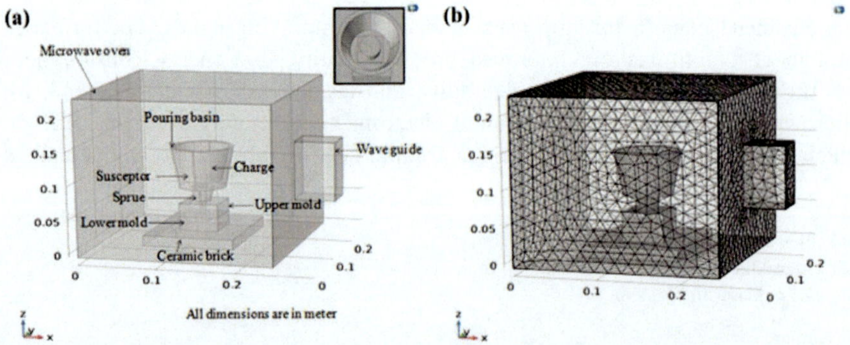

Fig. 3.1 **a** 3-Dimensional model of microwave melting experimental setup **b** Meshed domain of model

Table 3.1 Material properties used in simulation work [14, 15]

Materials	Density (kgm^{-3})	Specific heat (c_p) ($Jkg^{-1}K^{-1}$)	Thermal conductivity ($Wm^{-1}K^{-1}$)	Electric conductivity (Sm^{-1})	Permittivity	Magnetic permeability	Emissivity
Al	2700	$741.56 + 0.5071 \times T$	$238/(0.000011 \times T^{2.30} + (0.124/T))$	$(3.774 \times 10^7) / (1+0.0043 \times (T-300.15))$	1	1	0.19
Pb	11350	130	35	4.87×10^6	1	1	0.28
Mg*	1770	1000	96	1.087×10^7	1	1	0.2
Alumina	3900	900	27	0.27	8.9	1	–
Graphite	2100	830	470	1700	15	1	–

* COMSOL Multiphysics material library

3.3 Governing Equations

Microwave oven cavity is filled with only air at the ambient condition. Any chemical changes like - oxidation in charge materials have been neglected. The wall of the waveguide and microwave oven is made of copper metal. In the thermal analysis, the air domain and ceramic brick domain are not considered. A rectangular waveguide port is excited by transverse electric (TE10) field at a frequency of 2.45 GHz in the Z-direction only and waveguide has one input propagation mode.

The electromagnetic wave propagation inside the oven is governed by Maxwell's equations. The electric field **E** (V m^{-1}) inside the microwave oven is given by Eq. (3.1) [14]:

$$\nabla \times \mu_r{}^{-1}(\nabla \times E) - k_0^2\left(\varepsilon_r - \frac{j\sigma}{\omega\varepsilon_0}\right)E = 0 \tag{3.1}$$

where μ_r = relative permeability, σ = electrical conductivity, ε_r = relative permittivity, $j = \sqrt{(-1)}$, ω = angular frequency, ε_0 = vacuum permittivity ($8.85 \times 10^{-12} Fm^{-1}$), and k_0 denotes the wave number in free space.

The Microwave (MW) energy dissipates in the form of heat inside the material and can be expressed as [14]

$$\rho C_P \frac{\partial T}{\partial t} + Q_{Loss} = Q(r, T) + \nabla.(k\nabla T) \tag{3.2}$$

where ρ = density (kg m^{-3}), k = thermal conductivity ($Wm^{-2}K^{-1}$), C_P = specific heat capacity ($Jkg^{-1}K^{-1}$), Q_{Loss}= heat loss from the metal due to convection and radiation, and Q(r, T) = heat source that is microwave (MW) energy consumed by the material.

3.4 Result and Discussion

3.4.1 Electric Field Intensity

Figure 3.2a depicts the distribution of electric field intensity inside a microwave oven. The maximum and minimum electric field intensities are 6.39×10^4 V/m and 6×10^{-9} V/m, respectively, inside the microwave oven at 1400W power level. As the input power is proportional to the electric field, increasing the input power increases the electric field inside the oven (Fig. 3.2b).

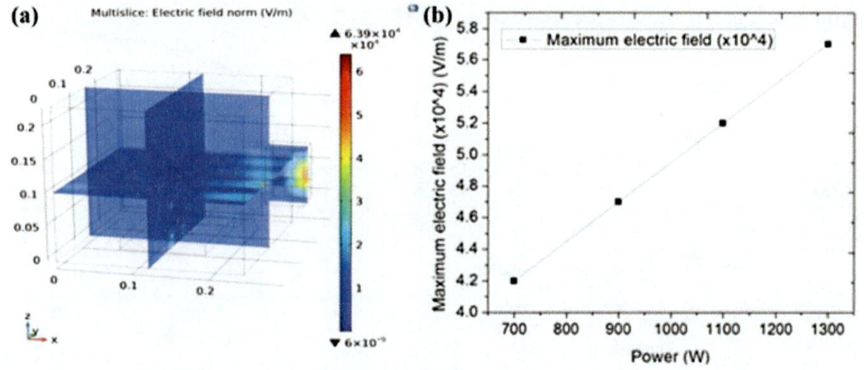

Fig. 3.2 **a** Distribution of electric field inside the cavity of microwave oven and **b** variation of maximum electric field inside the oven with power

3.4.2 Time-Temperature Profile of Melting of Metals

Bulk metal reflects microwave irradiations at room temperature. Bulk metals begin to absorb microwave irradiation once they reach critical temperature. Initially, the metal is heated with the help of a susceptor and heat transfer to the bulk metal takes place via conduction. As the bulk metal melts, it self-pours into the mold via the sprue. Figure 3.3a shows the comparison of the simulated time vs temperature profile of aluminum with the experimental data reported by Mishra and Sharma [12]. It is observed that the simulated temperature profile follows approximately the same trend as reported in literature. It is observed from Fig. 3.3b that lead starts to melt (327.5 °C) within 5 min at an input power of 1400W, while Mg alloy started to melt within 8 min at an input power of 1400W. The initial rate of heating in Mg alloy is high as compared to lead. Figure 3.3c shows the effect of casting volume on the temperature profile. Initially, an Mg alloy of constant volume V was used, and the volume was varied between 0.8 and 1.2 V. It was found that changing the volume of casting had no effect on the temperature profile (charge). In microwave heating, Chandrasekaran et al. reported that melting time is independent of material mass increase up to a certain limit [15].

3.5 Conclusions

COMSOL 5.6 was used to successfully simulate the melting of bulk metals (AA-7039, Mg alloy, and lead) at 1400W and 2.45GHz. The electric field inside the microwave oven increases as the input power level increases. At 1400W power level, the maximum and minimum electric field intensities inside the microwave oven are 6.39×10^4 V/m and 6×10^{-9} V/m, respectively. Aluminum alloy, magnesium alloy,

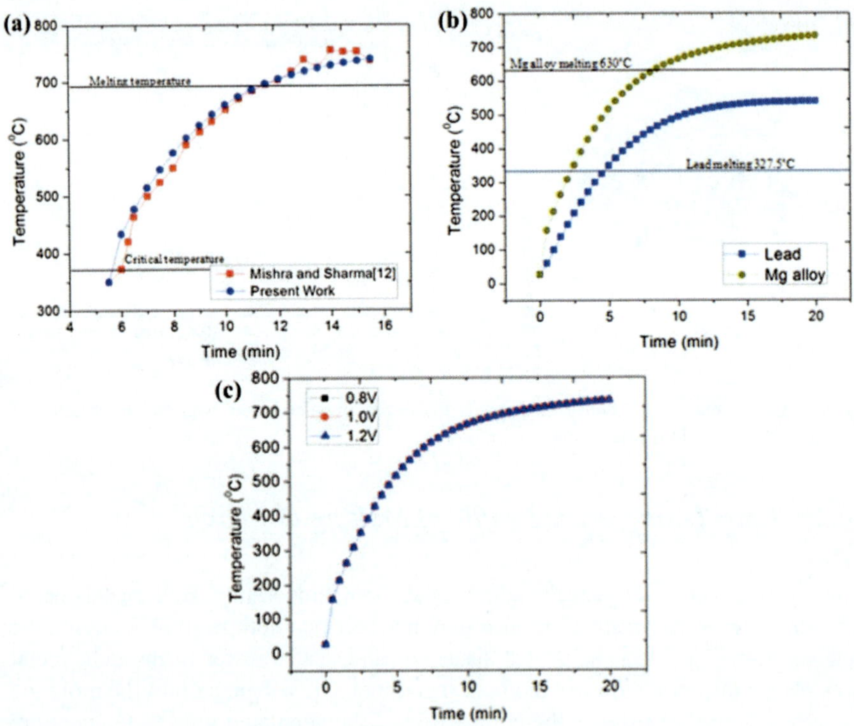

Fig. 3.3 **a** Time–Temperature graph of (a) AA-7039 **b** Lead and Magnesium alloy during microwave heating, **c** Effect of casting volume on the temperature profile with respect to time

and lead begin to melt in 11.5 minutes, 8 minutes, and 5 minutes, respectively. The temperature profile is unaffected by minor changes in charge volume.

Acknowledgments The authors sincerely acknowledge ISTEM for providing the facility to carry out the simulation work in COMSOL Multiphysics 5.6 software.

References

1. Jin, M., Tang, R., Ji, Y., Liu, F., Gao, L., Huisingh, D.: Impact of advanced manufacturing on sustainability: an overview of the special volume on advanced manufacturing for sustainability and low fossil carbon emissions. J. Clean. Prod. **161**, 69–74 (2017)
2. Gupta, M., Leong, E. W. W.: Microwaves and metals. John Wiley & Sons (2008).
3. Lauf, R.J., Bible, D.W., Johnson, A.C., Everleigh, C.A.: Two to 18 GHz broadband microwave heating systems. Microw. J. **36**(11), 24–30 (1993)
4. Thostenson, E.T., Chou, T.W.: Microwave processing: fundamentals and applications. Compos. Part A: Appl. Sci. Manuf. **30**(9), 1055–1071 (1999)
5. Roy, R., Agrawal, D., Cheng, J., Gedevanishvili, S.: Full sintering of powdered-metal bodies in a microwave field. Nature **399**(6737), 668 (1999)

6. Srinath, M.S., Sharma, A.K., Kumar, P.: A new approach to joining of bulk copper using microwave energy. Mater. & Des. **32**(5), 2685–2694 (2011)
7. Kumar, G., Sharma, A.K.: Role of dielectric fluid and concentrator material in microwave drilling of borosilicate glass. J. Manuf. Process. **33**, 184–193 (2018)
8. Gupta, D., Sharma, A.K.: Microwave cladding: a new approach in surface engineering. J. Manuf. Process. **16**(2), 176–182 (2014)
9. Singh, S., Gupta, D., Jain, V.: Microwave melting and processing of metal–ceramic composite castings. Proc. Inst. Mech. Eng., Part B: J. Eng. Manuf. **232**(7), 1235–1243 (2018)
10. Mishra, R.R., Sharma, A.K.: Microwave–material interaction phenomena: heating mechanisms, challenges and opportunities in material processing. Compos. Part A: Appl. Sci. Manuf. **81**, 78–97 (2016)
11. Singh, S., Gupta, D., Jain, V.: Novel microwave composite casting process: theory, feasibility and characterization. Mater. & Des. **111**, 51–59 (2016)
12. Mishra, R.R., Sharma, A.K.: On mechanism of in-situ microwave casting of aluminium alloy 7039 and cast microstructure. Mater. & Des. **112**, 97–106 (2016)
13. Warren, B., Awida, M.H., Fathy, A.E.: Microwave heating of metals. IET Microw.S, Antennas & Propag. **6**(2), 196–205 (2012)
14. Mishra, R.R., Sharma, A.K.: On melting characteristics of bulk Al-7039 alloy during in-situ microwave casting. Appl. Therm. Eng. **111**, 660–675 (2017)
15. Chandrasekaran, S., Basak, T., Ramanathan, S.: Experimental and theoretical investigation on microwave melting of metals. J. Mater. Process. Technol. **211**(3), 482–487 (2011)

Chapter 4
Layered Grain Growth of Aluminum Between MWCNT Layers During Sintering of Al–MWCNT Composites

Rahul Sharma, C. Sasikumar, and Jayashree Baral

Abstract The present study demonstrates how layered grain growth evolved in aluminum during sintering, of Al–MWCNT composites prepared through mechanical alloying. Mechanical alloying of aluminum and MWCNT produced a flake-shaped lamellar particle by flattening panicles and welding them together, and random welding produced lamellar granules. MWCNT lamellas limit the grain growth resulting in a 2D layer of aluminum between the MWCNTs during sintering. By adding more MWCNTs into aluminum, the interlayer thickness of layered composites was refined. The layer thickness of aluminum is 2.5 cm in Al–0.20 wt% MWCNT, 1.0 By adding more MWCNTs into aluminum, the interlayer thickness of layered composites was refined. The layer thickness of aluminum is 2.5 μm in Al–0.20 wt% MWCNT, 1.0 μm in Al–0.50 wt% MWCNT, and about 0.8 μm in Al–1.0 wt% MWCNT. In addition to the grain refinement and hardening effects of MWCNTs, the sintered composites exhibit increased hardness with increasing the CNT concentration in the composite.

Keywords MWCNT · Mechanical alloy · 2D aluminum layer · Microhardness

4.1 Introduction

CNTs (carbon nanotubes) were first discovered by Ijiima [1] and have been considered as an ideal reinforcement material for enhancing polymeric, ceramic, and metallic materials [2–6]. As CNTs have greater mechanical properties such as tensile strength of 150 GPa and modulus of elasticity of 1200 GPa, they are an exceptional reinforcement phase in high specific strength nanocomposites [7, 8]. The fabrication of metal matrix CNT composites is beneficial due to their high stiffness and strength, greater service temperature, and higher thermal and electrical conductivity. In recent research, a number of lightweight, high-strength composites have been

R. Sharma · C. Sasikumar (✉) · J. Baral
Department of Materials and Metallurgical Engineering MANIT Bhopal, Bhopal, MP, India
e-mail: csasimv@gmail.com

A. Tewari et al. (eds.), *Proceedings of the 3rd International Conference on Advances in Materials Processing: Challenges and Opportunities*, Springer Proceedings in Physics 293, https://doi.org/10.1007/978-981-99-1971-0_4

developed using aluminum and CNT composites to take advantage of aluminum's lightweight, and CNTs' high stiffness and strength. Composites containing aluminum and CNTs are produced using liquid-phase processing techniques. In liquid-phase processing, the particles or reinforcements are fused into the matrix, but the dispersion is poor due to the CNTs' small size and tendency to entangle or clustered because of van der Waals forces. In a few research studies, it has been found out that forming small amounts of Al_4C_3 at the interface increases interfacial bonding with aluminum. A powder metallurgical processing technique established the advantages of low-temperature solid state processing, better distribution, and avoiding the formation of high-temperature products such as Al_4C_3, which have a negative impact on mechanical properties.

4.2 Experimental Procedure

4.2.1 Materials

A high-purity acetylene gas (C2H2) is used to synthesize multi-wall carbon nanotubes (MWCNTs). Nickel powder with 99.5% purity and 300 mesh size is used as a catalyst. CNTs synthesized were mixed with 99% pure aluminum powder with an initial size of 200 mesh in order to make Al–CNT nanocomposites. A high pure argon gas is used for sintering studies of Al–CNT nanocomposites. In the present studies, all chemicals used were highly pure with trace amounts of impurities. The chemical composition of aluminum powder was determined using a wet chemical method using atomic absorption spectrometry.

Fe	Si	Mn	Pb	Cu	Ti	Al
0.5	0.1	0.02	0.03	0.02	0.03	Balanced

4.2.2 Methodology

A catalyst-assisted high-temperature thermal decomposition process is used for the synthesis of multiwall carbon nanotubes (MWCNT). A laboratory grade acetylene gas is used as a precursor and argon gas is used as a carrier gas. The thermal decomposition process is carried out in a tubular furnace made of alumina tubes. The C_2H_2/Ar gas mixtures are allowed flow at a rate of 50 sccm (standard cubic per minute). The ratio of C_2H_2/Ar gas mixtures is maintained as 1:2 for all the experiments and the decomposition temperature is maintained at 700 °C. A powder-grade nickel is used as catalyst and placed in a ceramic boat inside the ceramic tubes.

Fig. 4.1 Photographs of MWCNTs, milled powder composites, and sintered compacts in a range of 0, 0.2, 0.5, and 1.0, respectively

The MWCNT synthesized in the present study is mechanically alloyed with powder-grade aluminum particulates by using planetary ball milling techniques. A Fritch planetary ball mill Pulverisette 5.0 is used for mechanical alloying. Milling is performed with tungsten carbide balls of 10 mm diameter with a ball-to-powder ratio (BPR) of 10:1 at a speed of 250 rpm. There are no other additives or environment protectors used during milling. The milling experiments were carried out with 10 g of Al powder and 0.2, 0.5, and 1 weight percent of MWCNTs in a jar of a laboratory planetary ball mill. We milled for 2 hours with a sequence of 20-minute milling intervals followed by 30-minute breaks to avoid overheating. The milled powders are consolidated using 20 Tons hydraulic press to prepare green compacts of 10 mm in diameter and 5 mm in height. Compaction is conducted at a pressure of 70 kg/cm^2 in a hardened chrome steel die. The powder weight, powder compaction pressure, temperature, and all other conditions were maintained same for all the experiments. Five specimens were made for each composition. Al–CNT green compacts were sintered in a tube furnace under atmosphere-controlled conditions. The chamber is also purged with argon during sintering to prevent the composites from oxidizing. The sample was sintered for 4 h at 500 °C in the sintering experiment. The typical photographs of the MWCNT, milled powder composites, and the sintered compacts with 0, 0.2, 0.5, and 1.0 are shown in Fig. 4.1a, b, and c, respectively.

4.3 Results

4.3.1 SEM Analysis of Al–CNT Nanocomposite

This microstructure clearly shows Al grain growth as elongated plates with MWCNT layer forming at their interfaces, as can be seen in Fig. 4.2. Interestingly, Al and MWCNT phases segregated during sintering, resulting in a layered composite consisting of Al and CNTs. In addition, the AFM study also shows that Al–0.5 wt%

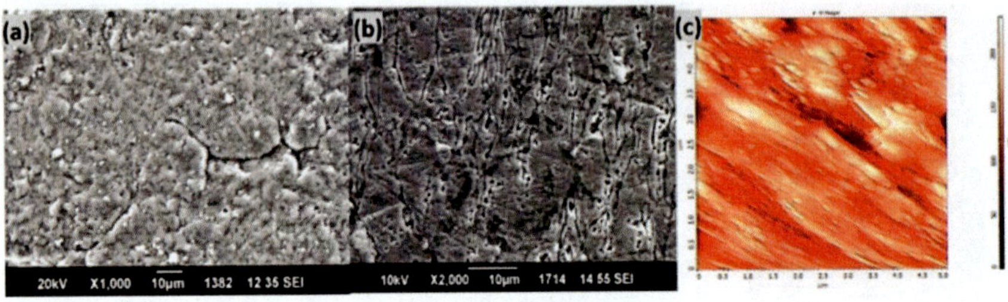

Fig. 4.2 **a**, **b** SEM images of Al–CNT composites prepared through mechanical alloying before and after sintering, 2 **c** AFM image of typical Al–CNT composite showing layered structure

CNT samples developed layered structure during sintering of mechanically alloyed materials. During sintering, this caused grain refinement, and their sizes are found to be about 1 μm in 0.2 wt% Al–CNT composites. The grain refinement is increased due to the addition of MWCNTs, and the grain size after sintering is approximately 0.8 μm in 0.5 wt% composite, and 0.5 μm in 1.0 wt% composite. The thickness of the carbon layers was found increasing with carbon percentage. It is also interesting to note that grain growth is decreased to a larger extent in Al–CNT nanocomposites. Grain growth inhibition by the addition of WC, VC, and other alloying elements is reported elsewhere. However, the role of CNT in restricting the grain growth during sintering is less reported. The composite structure also showed particulates dispersed inside the matrix phase.

4.3.2 Hardness

Micro Vickers hardness value is calculated using micro Vicker hardness tester with a diamond indenter. Hardness of the sample was found to be increasing from pure aluminum compact to increasing wt% of CNT in Al composite. For pure aluminum compact hardness value was found 38.23 $HV_{0.25}$ and varying the CNT content in pure aluminum by 0.2%, 0.5%, and 1%, the hardness value for the samples were found 49.45$HV_{0.025}$, 58.32$HV_{0.025}$, and 78.12$HV_{0.025}$, respectively. The highest value of hardness was found at 1wt% of CNT content in aluminium (Fig. 4.3).

4.4 Conclusion

- Directionally oriented grain growth during sintering is successfully demonstrated by adding MWCNTs into aluminum in the present work. The dispersion of MWCNTs into aluminum powders through mechanical alloying gives way to directional grain growth in aluminum during sintering.

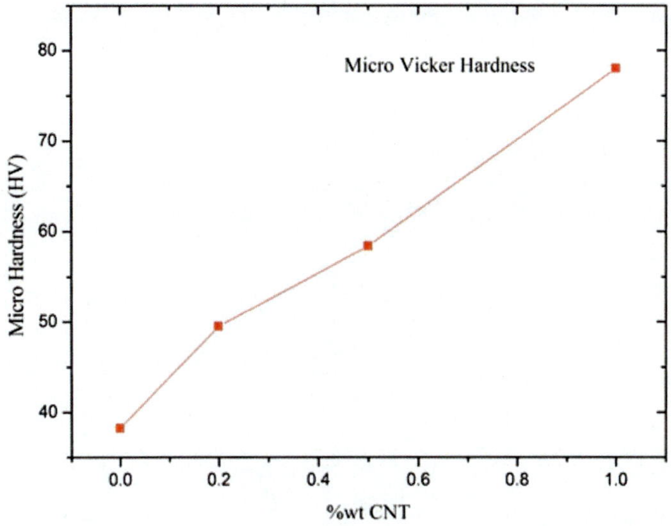

Fig. 4.3 Microhardness value of Al–MWCNT composite with different wt% (0, 0.2, 0.5, and 1.0) of CNT

- In mechanical alloying, the aluminum and CNT-rich phases were micro-separated into flake-like/lamellar structures, and the grain growth was restricted within the CNT layers during sintering.
- The hardness of the composites was increased and a maximum hardness was achieved at 78.12 HV in Al–CNT composites with 1 wt% of MWCNTs. Composites of Al–CNT demonstrate a 110% increase in hardness.

References

1. Iijima, S.: Helical microtubules of graphitic carbon. Nature **354**, 56–58 (1991)
2. Tjong, S.C.: Recent progress in the development and properties of novel metal matrix nanocomposites reinforced with carbon nanotubes and graphene nanosheets. Mater. Sci. Eng. R Rep. **74**, 281–350 (2013)
3. M. Estili, Y. Sakka, Recent advances in understanding the reinforcing ability and mechanism of carbon nanotubes in ceramic matrix composites, Sci. Technol. Adv. Mater. (2016). 4.
4. Liu, Y., Kumar, S.: Polymer/carbon nanotube nanocomposite fibersea review. ACS Appl. Mater. Interfaces **6**, 6069–6087 (2014)
5. Treacy, M.J., Ebbesen, T., Gibson, J.: Exceptionally high Young's modulus observed for individual carbon nanotubes. Nature **381**, 678–68 (1996)
6. Esawi, A., Morsi, K.: Dispersion of carbon nanotubes (CNTs) in aluminum powder. Compos. Part A **38**, 646–650 (2007)
7. George, R., Kashyap, K.T., Rahul, R., Yamdagni, S.: Strengthening in carbon nanotube/aluminium (CNT/Al) composites. Scripta Mater. **53**, 1159–1163 (2005)
8. W. De Heer, Nanotubes and the pursuit of applications, MRS Bull. (2004) 281.

Chapter 5
An Attempt to Develop an Organ-on-a-Chip Using 3D Printing Technology for in Vitro Drug Testing

Botcha Appalanaidu, Tara Chand Kumar Maurya, and Maran Rajakumaran

Abstract A medical drug, before entering the market, has to pass several clinical tests such as cell culture and animal tests. Cell culture tests are not reliable and not all animal modal suits humans. In addition, both the methods take several years and are expensive. In order to mitigate this issue, a new technology is emerging in the field of pharmaceutical industry for drug testing, i.e., organ-on-a-chip. The present research article aims at the fabrication of an organ-on-a-chip using 3D printing technology. A desired model has been designed in solidworks software and the supported file has been given to a commercialized 3D printer. A mold has been printed with stereolithographic technology. With the help of ultrasonication, supporting material is cleaned. A Polydimethylsiloxane (PDMS) solution, is used to prepare a chip base with a desired skeleton of organ. A desiccator has been used for degasification to eliminate any presence of gas voids. FE-SEM technique was employed to characterize channel and hole dimensions, and it has been found that average dimensional errors fall under $\pm 5\%$ which are promising results. Flexible silica microtubes have been attached at inlet and outports. At last, the chips have been tested for the feasibility of seamless microfluids flow with the help of PIV experimental setup, which is augmented with a programmable fluid flow regulator.

Keywords Microfluidic device · Lab-on-a-chip · Pharmaceutics

B. Appalanaidu (✉)
GITAM (Deemed to be University), Visakhapatnam, Andhra Pradesh, India
e-mail: banaidu318@gmail.com

T. C. K. Maurya · M. Rajakumaran
Indian Institute of Technology, Roorkee, Uttarakhand, India

A. Tewari et al. (eds.), *Proceedings of the 3rd International Conference on Advances in Materials Processing: Challenges and Opportunities*, Springer Proceedings in Physics 293, https://doi.org/10.1007/978-981-99-1971-0_5

5.1 Introduction

5.1.1 Organ-on-a-Chip

In the current scenario, one of the major global issues in the field of the pharmaceutical industry is drug testing. The average money invested in developing one successful drug is about 1.5 billion US dollars for at least 12 years [1]. There are multiple reasons involved. One of them is the current approaches available for conducting pre-clinical tests. There are two basic types of pre-clinical drug testing methods, those are in vitro models and in vivo models. In vitro models, cell culture and animal testing are more prominent [2]. In cell culture, live human body cells are collected from respective organs and placed in a glass dish to conduct a drug test. Cell culture produces poor results due to the lack of a native environment for cells [3]. The other one is animal testing. Drug testing on animals provides useful information and leads to explore more about biology. However, animal models failed to predict what will happen in human when there are treated with a particular drug. It is mentioned that some testings are very effective in vitro models but not in vivo models. Thus, better approaches and tools are needed in which human cells should participate in drug testing in the same complex and dynamic way as in our bodies. In addition, the method should provide results at lower cost at rapid time. There comes organ-on-a-chip. The organ-on-a-chip is a tiny device made of glass and/or Polydimethylsiloxane (PDMS). The functional unit that represents the respective organ is recreated on a chip. In addition, the cells in this unit experience the same dynamic mechanical strains and fluid motions as in the human body.

5.2 Organ-on-a-Chip Models for Various Organs

In recent times, extensive research is going on to develop organ-on-a-chip models for respective human organs such as kidney, liver, heart, lung, vessels, etc., to provide better possible results in vitro testing. Weinberg et al. [4] reported the development of kidney-on-a-chip. This chip has been fabricated with two microfabricated layers which are separated by a membrane. A lung-on-a-chip model was developed by Huh et al. [5] with the help of two-layered channels and a micro membrane (thin, porous, and flexible) made up of PDMS material to investigate airborne diseases. Grosberg et al. [6] developed a heart-on-a-chip using a muscular thin film technique. In order to monitor the systolic and diastolic cycles a digital video-assisted external field simulator was used. Zhang et al. [7] reported a microfluidic device for vascular phenomenon using conventional micromachining. However, this technique faced challenges for cell functioning and fluid flow behavior during drug testing. Lee et al. [8] developed a microfluidic device that mimics the human liver with microchannel, microholes, and membranes. A photolithography was used for the fabrication of cell loading and fluid flow.

5.3 Fabrication Processes to Develop Organ-On-A-Chip Models

Till several fabrication methods have been reported for the fabrication of organ-on-a-chip devices. Zheng et al. [7] developed a microfluidic device for vascular phenomenon using conventional micromachining. Researchers have attempted to develop microfluidic devices using electrochemical discharge machining process on glass material [9–11]. Lee et al. [8] reported the stereolithography bioprinting technique for fabrication of biochips. In this process, a 2D structure of organ is printed layer by layer using bio-ink and then solidified selectively by photopolymerization. A paper-based inkjet bioprinting has been introduced by Boland et al. [12], which involves droplet-wise formation of bio-ink to form a 2D pattern. Extrusion-based bioprinting has been introduced by Zein et al. [13], which involves scaffold fabrication using extrusion method.

The above methods involve sophisticated lab environment and expensive equipment, which is not academic research-friendly. Therefore, an alternative choice to fabricate microchips with rapid speed, low cost, and a scale relevant to both the cell and its environment is needed. In the present proposed research, a microfluidic device has been developed using stereolithographic technique. A mold has been prepared using a multijet printer and a pattern has been prepared with the help of PDMS solution. Ultrasonication and degasification have been conducted to improve the microfeatures' dimensional accuracy. With the help of cover glass and microsilica rubber tubes, an organ-on-a-chip has been developed. The chip has been tested for microfluid flow with a programmable fluid flow regulator.

5.4 Material and Methods

In the present research, an organ-on-a-chip has been fabricated using the Indian Institute of Technology, Roorkee, Tinkering Lab 3D printing technology facilities. A commercial solidworks software has been used for designing a chip. A mold has been prepared by the industrial 3D printer (Make: 3D Systems, Model: MJP3600) with XHD mode, which has accuracy in layer to layer (height) 16 μm, and dimension (length and width) 30 μm (750 $\times$ 750 $\times$ 1600 dpi). visiJet M_3 crystal material is being used for mold preparation and wax is used for support material. Ultrasonicator (Make: Citizen, Model: FB15046) is used for cleaning the mold for foreign particles. Polydimethylsiloxane (PDMS) is used for microfluidic device fabrication. For degasification, a desiccator has been employed. For dimensional characterizations, FE-SEM (make: Carl Zeiss, model: GeminiSEM 300) having magnification 100–100,000 (BSE) is used. Microsilica tubes are used to interconnect the microchannels and also for inlet and outlet ways.

5.5 Development of an Organ-On-A-Chip

Initially, a 3D CAD drawing is designed in a commercial solid works software for desired dimensions of the skeleton of an organ (Shown in Fig. 5.1a). The same file given to multijet printer with STL format. This industrial printer works with visiJet M_3 crystal as the main material and wax as the supporting material. The 3D printer develops a mold as per the design with an accuracy of 16 μm layer to layer thickness as shown in Fig. 5.1b. During the mold development, sometimes, wax deposition occurs on mold surface. Before, it is used for molding, the wax must be entirely removed. Otherwise, insufficient cleaning of surface leads to the rise in the difficulty in the solidification of PDMS material in a controlled environment. There are three mediums to remove the wax such as oil bath, deionized water, and soap solution. After cleaning with any one of the mediums, the mold undergoes 30 min of ultrasonication to clean other residuals.

Further, a mixture of PDMS with a monomer ratio of 10:1 has been prepared. It has been poured on the mold and placed in the desiccator for the process of degasification. The same is placed in a furnace for 80 min at 75 °C to get the desired solidification. The required microfluidic pattern has been developed as shown in Fig. 5.2a. A test has been conducted with FE-SEM to assure the channel dimension accuracy and surface property. It has been found that the average error in the dimension falls under ± 5%, which is promised result in the experiments. A borosilicate cover slide is used to seal the microfluidic pattern. The assembly is exposed to ionized air during plasma treatment. The ionized air activates the surface of the PDMS channel and covers the slide and then is allowed to stick together by manual pressure. Flexible microsilica tubes are used to insert via microholes for inlet and outlet ways to develop an organ-on-a-chip (Figs. 5.2b and 5.3).

At last, the developed chip has been tested for feasibility of seamless microfluids flow with the help of a PIV experimental setup, which is augmented with a programmable fluid flow regulator.

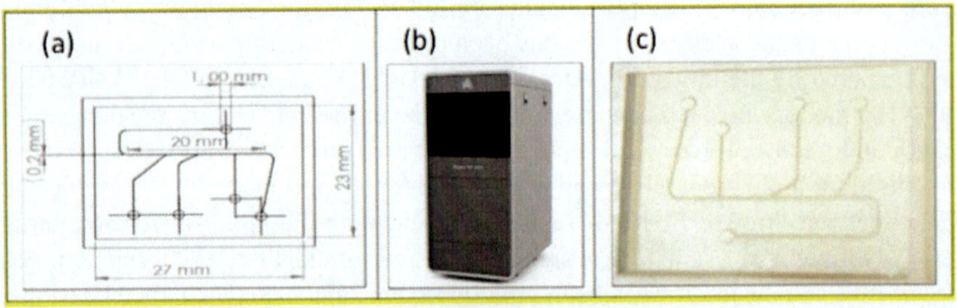

Fig. 5.1 **a** CAD drawing of organ-on-a-chip **b** Multijet printer **c** Mold for organ-on-a-chip

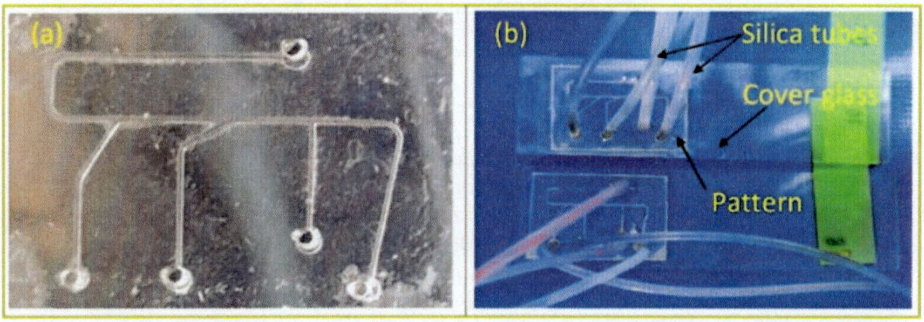

Fig. 5.2 **a** PDMS pattern for an organ-on-a-chip **b** Assembling model of an organ-on-a-chip

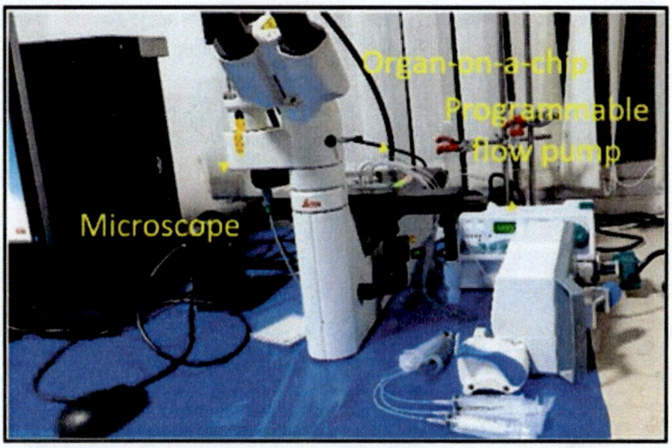

Fig. 5.3 Testing of an organ-on-a-chip for seamless fluid flow

5.6 Conclusions

In the present investigation, an attempt has been made for the development of an organ-on-a-chip using 3D printing technology. Conclusions are drawn as follows:

- This method can be an alternative choice to fabricate microchips with rapid speed, low cost, and a scale relevant to both the cell and its environment.
- Using these devices, a strong database can be maintained such as real-time monitoring of changes in cell damage, cell death, and cell response to the drug.
- This method provides flexibility design changes in rapid time and supports the feature complexity of the microfluidic pattern.

References

1. Esch, M.B., King, T.L., Shuler, M.L.: The role of body-on-a-chip devices in drug and toxicity studies. Annu. Rev. Biomed. Eng. **13**(1), 55–72 (2011)
2. Ghaemmaghami, A.M., Hancock, M.J., Harrington, H., Kaji, H., Khademhosseini, A.: Biomimetic tissues on a chip for drug discovery. Drug Discov. Today **17**(3–4), 173–181 (2012)
3. Sung, J.H., Kam, C., Shuler, M.L.: A microfluidic device for a pharmacokinetic–pharmacodynamic (PK–PD) model on a chip. Lab Chip **10**(4), 446–455 (2010)
4. Weinberg, E., Kaazempur, M. M., Borenstein, J.: Int. J. Artif. Organs. **31**, 508–514.-14 (2008).
5. Huh, D., Matthews, B. D., Mammoto, A., Montoya, Z. M., Hong, Y. H., Donald, E. I.: Sci. **328**, 1662–1668.-20 (2010).
6. Grosberg, A., Alford, P.W., McCain, M.L., Parker, K.K.: Ensembles of engineered cardiac tissues for physiological and pharmacological study: heart on a chip. Lab Chip **11**(24), 4165–4173 (2011)
7. Zheng, Y., Chen, J., Craven, M., Choi, N.W., Totorica, S., Diaz-Santana, A., Kermani, P., Hempstead, B., Fischbach-Teschl, C., López, J.A., Stroock, A.D.: In vitro microvessels for the study of angiogenesis and thrombosis. Proc. Natl. Acad. Sci. **109**(24), 9342–9347 (2012)
8. Lee, M. P., Cooper, G. J., Hinkley, T., Gibson, G. M., Padgett, M. J., Cronin, L.: Sci. Rep. **5**, 9875.-63 (2015).
9. Appalanaidu, B., Dvivedi, A.: On controlling of gas film shape in electrochemical discharge machining process for fabrication of elliptical holes. Mater. Manuf. Processes **36**(5), 558–571 (2021)
10. Singh, T., Appalanaidu, B., Dvivedi, A.: Improvement in energy channelization behaviour during micro hole formation in Y-SZ ceramic with magnetic field assisted ECSM process. Measurement **194**, 111079 (2022)
11. Appalanaidu, B., Dvivedi, A.: On the use of sacrificial layer in ECDM process for form accuracy. J. Manuf. Process. **79**, 219–232 (2022)
12. Boland, T., Wilson, W. C., Xu, T.: U.S. Patent No 7051654 B2. US Patent Office, Washington, DC, USA,-64 (2003).
13. Zein, I., Hutmacher, D. W., Tan, K. C., Teoh, S. H.: Biomaterials. **23**, 1169–1185.-68 (2002).

Chapter 6
Investigation on Through Mask Electrochemical Surface Texturing Using Additive Manufactured Mask

Maran Rajakumaran and Akshay Dvivedi

Abstract Surface texturing plays a critical role in improving the functionality of the part surface in terms of biological performance, hydrophobicity, friction coefficient, etc. In comparison to other methods, through mask electrochemical machining (TMECM) is a versatile surface texturing process for machining difficult-to-machine metals. However, the conventionally employed photolithographic masking in the TMECM process is expensive, time-consuming, and lacks mask re-usability. Hence, a novel technique has been identified wherein the workpiece (anode) is masked using patterned insulation printed using additive manufacturing technique. The main advantage of this technique is that the mask fabrication is a one-time process, and the same can be reused further for subsequent machining processes. The competence of the developed technique is experimentally verified by machining micro-dimples and various other features on SS304 material.

Keywords TMECM · Surface texturing · Additive manufacturing

6.1 Introduction

The advent of new technology obligates the invention and usage of metals with enhanced properties such as high strength-to-weight ratio, wear resistance, corrosion resistance, etc. But metals with such superior properties make them challenging to machine using the traditional machining processes [1]. In addition, machining of complex and repetitive features demands the substitution of traditional with non-traditional processes under certain case-specific circumstances [2, 3]. But before choosing the apt non-traditional process, several factors, such as

M. Rajakumaran (✉) · A. Dvivedi
Department of Mechanical and Industrial Engineering, Indian Institute of Technology Roorkee, Roorkee, Uttarakhand 247667, India
e-mail: mr@me.iitr.ac.in

A. Dvivedi
e-mail: akshaydvivedi@gmail.com

A. Tewari et al. (eds.), *Proceedings of the 3rd International Conference on Advances in Materials Processing: Challenges and Opportunities*, Springer Proceedings in Physics 293, https://doi.org/10.1007/978-981-99-1971-0_6

machining cost, time, efficiency, environmental impact, surface quality, part application, post-processing, etc., needs to be considered. By considering the above factors for texturing on metallic surfaces, through mask electrochemical machining (TMECM) process significantly outweighs other non-traditional processes. For instance, machining a single hole or an array of 1000 holes would consume the same amount of time with TMECM process. Also, the lack of heat-affected zone (HAZ), thermal stress, and recast layer makes ECM a versatile process [4]. The presence of the above-mentioned factors in the case of thermal-based processes such as electric discharge machining (EDM), laser beam machining (LBM), ion beam machining (IBM), etc., are detrimental to the part life and requires further post-processing. Subsequently, ECM is the preferred solution even for post-processing parts produced by thermal-based processes.

The role of surface texturing and its impact on the performance of a part in terms of wettability, friction coefficient, reflectivity, etc., has been well established by several researchers [5–7]. Though surface texturing can be performed using multiple techniques such as photochemical machining (PCM), LBM, TMECM, etc., TMECM has been the most sought-after process due to its aforementioned advantages. In TMECM process, surface texturing is largely carried out by masking the workpiece using photolithographic technique. Despite photolithography being an effective masking technique, the sequence of processing steps and the lack of mask re-usability are its limitations. The inability to reuse the mask significantly increases the cost and time of subsequent machining operations. In addition, the difficulty in adjusting the thickness of the mask may lead to hump formation and affect the functionality of the textured part. To overcome the above-mentioned issues, several researchers [8–10] utilized polydimethoxylsiloxane (PDMS) mask, maskless or pattern transfer masking, etc. In the present work, a novel way of mask fabrication using additive manufacturing has been employed as an alternative to the photolithographic masking technique. The effectiveness of additive manufactured mask has been studied by varying the mask thickness and duty cycle and evaluating the undercut of the machined blind hole. Undercut is defined as the difference between the radius of the machined hole and the mask hole.

6.2 Experimental Details

TMECM is a variant of ECM process wherein the tool and masked workpiece are held stationary and anodic dissolution occurs through the exposed regions of the patterned mask when connected to the DC power supply. The use of a pulsed power supply in TMECM process is to ensure that ample time is available for the pump to get rid of the sludge and bubbles from the machining zone. The presence of the same reduces the material removal rate and causes non-uniform machining. The schematic of TMECM process is shown in Fig. 6.1. The machining parameters for TMECM process are given in Table 6.1. The design of experiments was carried out using Taguchi L9 orthogonal array with two factors and three levels.

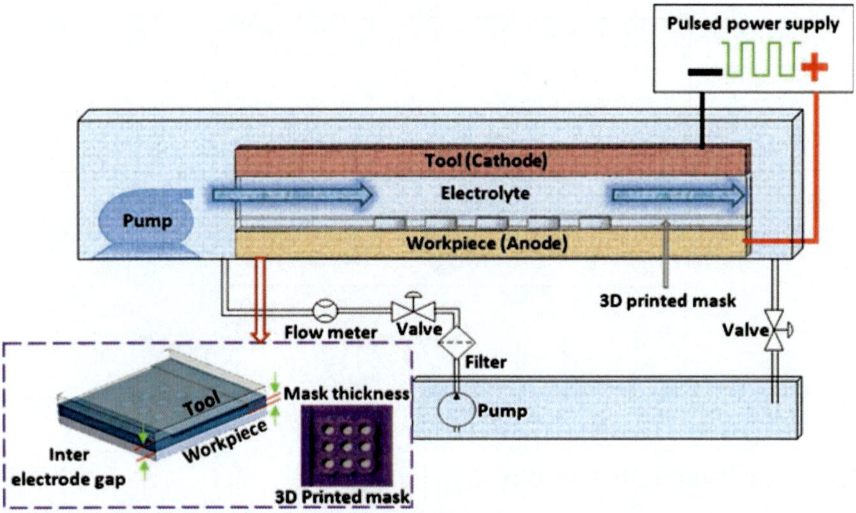

Fig. 6.1 Schematic of TMECM process

Table 6.1 Experimental conditions

Parameters	Values
Electrolyte concentration	15 wt% $NaNO_3$
Electrolyte temperature	25 °C
Workpiece material/Tool material	SS 304
Workpiece thickness	100 μm
Electrolyte velocity	2.5 m/s
Diameter of mask hole	2 mm
Machining time	1 min
Input voltage	12 V
Pulse frequency	1 kHz
Mask thickness	1 mm, 1.25 mm, 1.5 mm
Inter electrode gap	Mask thickness + 0.3 mm
Duty cycle	25%, 50%, 75%

In the present work, Prusa SL1S, a stereolithographic additive manufacturing printer was utilized for fabricating the patterned mask for TMECM process. The mask material was made from a UV-sensitive resin that solidifies by photopolymerization. The CAD model (.stl format) of the patterned mask was fed into the printer after slicing using the open-source PrusaSlicer 2.5.0 software. The patterned mask comprises of an array of nine holes, each with a radius of 1 mm. Masks with three different thicknesses (1 mm, 1.25 mm, and 1.5 mm) were used in the present study.

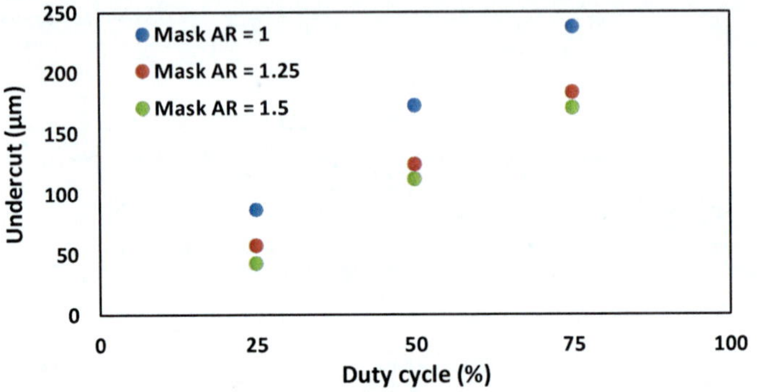

Fig. 6.2 Effect of mask aspect ratio and duty cycle on undercut

6.3 Results and Discussion

In the present study, the feasibility of utilizing 3D printed masks for TMECM process has been explored. Although researchers working in the domain of TMECM process have numerically simulated the effect of mask thickness >1 mm, experimental validation was not attempted due to the practical limitations of masking using photolithography. The use of 3D printing in the present study has made it possible to fabricate masks of thickness >1 mm. The effect of mask aspect ratio (AR) and duty cycle on undercut is shown in Fig. 6.2. For a set value of duty cycle, it is observed that an increase in mask AR from 1 to 1.5 decreases the undercut, whereas an increase in duty cycle from 25% to 75% increases the undercut. Increasing the mask thickness increases the inter-electrode gap leading to reduced current density and hence less material removal rate (MRR). Thus, the observed reduction in undercut with increased mask AR is the result of reduced MRR. Also, the increase in duty cycle from 25% to 75% increases the MRR and undercut due to the increased pulse on time. The reverse is also likely to occur if proper flushing is not provided to evacuate the sludge and bubbles from the machining zone. From the SEM images shown in Fig. 6.3, it is evident that the use of high aspect ratio masks has resulted in machining of features without hump formation. The notable observation from the present study is that even though high aspect ratio masks are beneficial, increasing the mask thickness beyond a certain value is detrimental to the efficiency of the process.

6.4 Conclusion

In the present study, the viability of using 3D printed masks in TMECM process was carried out successfully by machining dimples and alphabets on SS304 work material. Two parameters, namely, mask thickness and duty cycle were varied, and their effect on undercut was studied. The present study is not entire in itself and

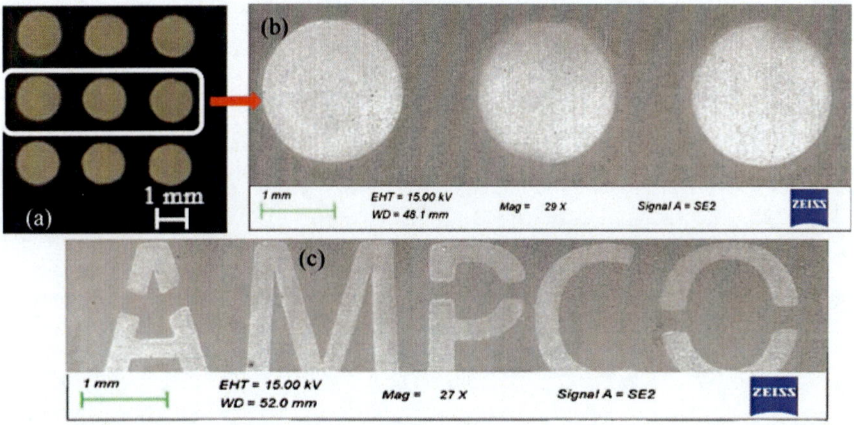

Fig. 6.3 Microscopic image of **a** array of micro-dimples, SEM image of **b** micro-dimples and, **c** alphabets textured using 3D printed mask and TMECM process

the effect of other parameters and responses needs to be studied for a thorough understanding. But nonetheless, it has become evident that 3D printed masks can be effectively used as a mask in TMECM process.

References

1. Kar, S., Patowari, P.K.: Parametric optimization of micro-electrical discharge drilling on titanium. In: Advances in Micro and Nano Manufacturing and Surface Engineering: Proceedings of AIMTDR 2018, pp. 201–210. Springer, Singapore (2019)
2. El-Hofy, H.: Vibration-assisted electrochemical machining: a review. Int. J. Adv. Manuf. Technol. **105**(1–4), 579–593 (2019)
3. Hu, X., Zhu, D., Li, J., Gu, Z.: Flow field research on electrochemical machining with gas film insulation. J. Mater. Process. Technol. **267**, 247–256 (2019)
4. Patel, D.S., Agrawal, V., Ramkumar, J., Jain, V.K., Singh, G.: Micro-texturing on free-form surfaces using flexible-electrode through-mask electrochemical micromachining. J. Mater. Process. Technol. **282**, 116644 (2020)
5. Byun, J.W., Shin, H.S., Kwon, M.H., Kim, B.H., Chu, C.N.: Surface texturing by micro ECM for friction reduction. Int. J. Precis. Eng. Manuf. **11**(5), 747–753 (2010)
6. Marmur, A.: Super-hydrophobicity fundamentals: implications to biofouling prevention. Biofouling **22**(02), 107–115 (2006)
7. Kubiak, K.J., Wilson, M.C.T., Mathia, T.G., Carval, P.: Wettability versus roughness of engineering surfaces. Wear **271**(3–4), 523–528 (2011)
8. Chen, X., Qu, N., Li, H., Xu, Z.: Electrochemical micromachining of micro-dimple arrays using a polydimethylsiloxane (PDMS) mask. J. Mater. Process. Technol. **229**, 102–110 (2016)
9. Costa, H.L., Hutchings, I.M.: Development of a maskless electrochemical texturing method. J. Mater. Process. Technol. **209**(8), 3869–3878 (2009)
10. Kunar, S., Kumar, R., Varaprasad, B., Sree, S.R., Murthy, K.V.S.R., Reddy, M.S.: Micropatterning using maskless electrochemical micromachining. Mater. Today Proc. **65**, 3273–3277 (2022)

Chapter 7
Study of the Microstructure and Mechanical Behavior of TIG Weldments Between AISI 304L and AISI 430 Stainless Steels

Markush Bakhla, Ajaykumar Udayraj Yadav, Binod Kumar, and M. Sambasiva Rao

Abstract The present study aims to determine the effect of ER347 welding electrode and the behavior of similar and dissimilar weldments of austenitic AISI 304L steel and ferritic AISI 430 stainless steels, which were prepared by Gas Tungsten Arc Welding (GTAW) process. The effect of alloying elements on the microstructural and mechanical properties of the weldments was investigated. The variation in microstructure at heat-affected zone (HAZ) and weld regions were analyzed. The similar 304L–304L weldment is characterized by the presence of delta ferrite stringer in the transition region, whereas, in the dissimilar 304L-430 weldment, significant grain coarsening in the transition region of ferrite with the 200–240 μm diameter grain size, has been noticed. Moreover, the grain boundaries between the weld and the ferritic regions are found to be decorated with carbides. Tensile test and microhardness were performed to evaluate the mechanical properties of the specimens. The results revealed that the 304L–304L weldments provided higher tensile strength and % elongation but lower hardness values among all the different weldments.

Keywords AISI 304L · AISI 430 · TIG-Welding

7.1 Introduction

Dissimilar welding provides extensive grades of engineering materials for their use in high-temperature structural applications. Dissimilar welding of the thin section was involved in joining two or more different alloys by gas tungsten arc welding (GTAW)

M. Bakhla · A. U. Yadav · B. Kumar (✉)
Materials and Metallurgical Engineering Department, NIAMT, Hatia, Ranchi, India
e-mail: binodkumar@niamt.ac.in

M. Sambasiva Rao
Foundry Forge Technology Department, National Institute of Advanced Manufacturing Technology (Formerly-NIFFT), Ranchi 834003, India

A. Tewari et al. (eds.), *Proceedings of the 3rd International Conference on Advances in Materials Processing: Challenges and Opportunities*, Springer Proceedings in Physics 293, https://doi.org/10.1007/978-981-99-1971-0_7

which is an inexpensive and productive way to connect these plates for several applications [1–3]. These procedures were preferably proposed for the production of heat exchanger of boiler which needs to be at service from low to high temperatures. The GTAW process was extensively employed for joining high to low carbon content of ferritic to austenitic stainless steel. During welding of austenitic and ferritic stainless steel, Nickel (Ni) plays an important role in fusion and heat-affected zones (HAZ) which affect the final microstructural and mechanical properties [4, 5]. Whereas, after joining, ferritic stainless steel deals with the issue of grain coarsening between weld region and HAZ region. Which in turn, reduces the toughness and ductility significantly in the welded region of ferritic stainless steel [5, 6]. Therefore, transition welding needs to be explored in such areas of applications, where weldments between austenitic–ferritic stainless steel have to withstand in transition temperature region. Therefore, the present study focuses on the behavior of similar and dissimilar weldments of austenitic AISI 304L steel and ferritic AISI 430 stainless steels, prepared by using Gas Tungsten Arc Welding (GTAW) process.

7.2 Experimental Methods

As-received austenitic AISI 304L and ferritic AISI 430 stainless steel plates of 4 mm thickness with the dimension of 155 mm X 100 mm were joined by GTAW welding using welding electrode ER347. Three sets of plates were developed similar 430–430, dissimilar 430-304L, and similar 304L-304L. Table 7.1 shows welding parameters and Table 7.2 shows the chemical composition analysis by wet chemical analysis.

The successful weld samples were cut into $10 \times 10 \times 4$ mm dimensions by using wire-cut EDM from the welded zone. These samples were polished with different grades of emery paper and cloth using 0.25 μm diamond paste. Two different reagents were used for etching: Villella's reagent (100 ml ethanol + 5 ml HCl + 1 gm picric acid) for AISI 430 stainless steel and a solution of 10 ml HNO_3 + 20 ml HCl + 30 ml

Table 7.1 Welding parameters of GTAW process

Shielding gas	Voltage (V)	Current (A)	Electrode diameters (mm)	Welding Speed (m/mm)	Gas flow rate (L/min)
Pure-Ar	22.8	180–200	1.2	6.0	6.5
Pure-Ar	22.9	180–200	1.2	6.0	6.5

Table 7.2 Chemical composition of base metals and welding electrode

Material	C	S	P	Mn	Si	Cr	Ni	Mo	Fe
AISI304L	0.028	0.008	0.031	1.35	0.39	18.05	8.02	–	Bal
AISI430	0.060	0.008	0.031	0.50	0.35	16.01	–	–	Bal
ER347	0.050	0.02	0.015	1.26	0.33	20.05	10.18	0.12	Bal

water from ASTM-E407-07 standard for weldments and AISI 304L stainless steel. The micro-hardness of similar and dissimilar weldment specimens was measured using the 300 g load for a dwell time of 15 s. For tensile testing, three rectangular transverse tensile specimens of each weldment were prepared as per the ASTM E08 standard.

7.3 Results and Discussion

The macrostructure of similar weldments between 304L-304L is shown in Fig. 7.1A. The different structures of transition region using the GTAW process are also evident in Fig. 7.1A. Similarly, such transition regions were also observed in dissimilar 304L-430 and similar 430–430 weldments as shown in Fig. 7.2. It could be evident that no visual and metallurgical defects are present in all three weldments. Optical microstructures of all-transition regions of similar and dissimilar weldment plates illustrate the phase transformation as determined by their chemical composition.

Initially, the as-received austenitic stainless steel reveals nearly equiaxed polygonal grains with annealing twins [1]. After welding of similar austenitic stainless steel, the development of delta ferrite stringer [2] in the transition region of their load axis occurs which comes through the cold work as shown in Fig. 7.1a and c. Moreover, in the whole welded region, equivalent formation of austenite and delta ferrite phase occurs with coaxial and columnar structure [1, 2] as exhibited in Fig. 7.1b.

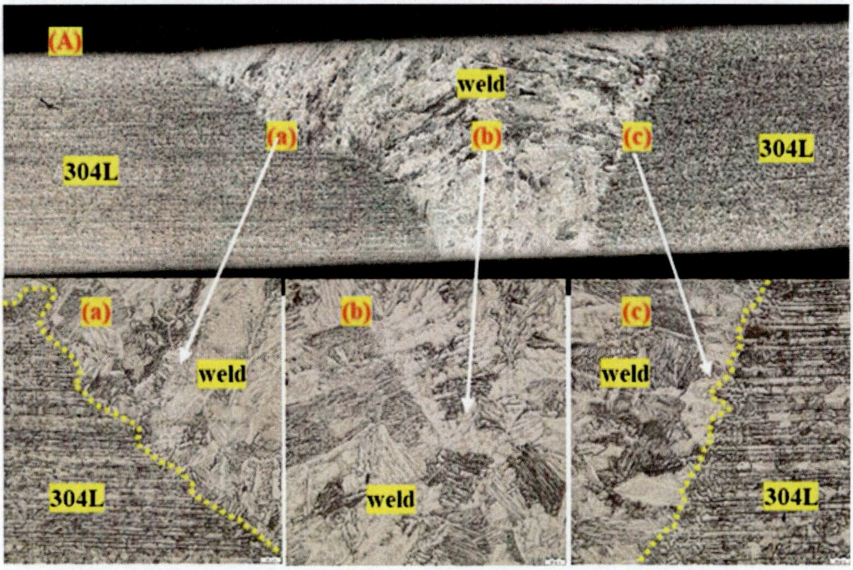

Fig. 7.1 Macrostructure and microstructure of similar weldments between AISI 304L-AISI 304L provides the distinguished region with their phase discrimination

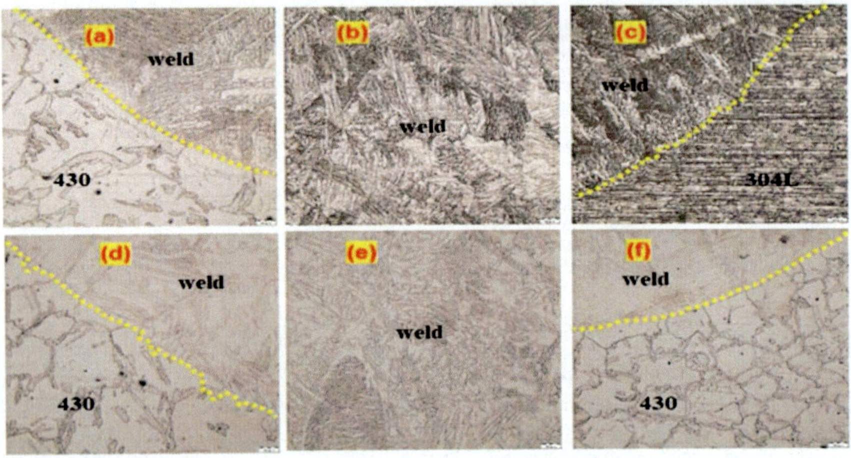

Fig. 7.2 **a–c** Microstructure of the dissimilar weldments shows the HAZ and weld zone between 304L-430. **d–f** Similar weldments show the HAZ and weld zone between 430–430

Microstructure of dissimilar weldments shows the significant grain coarsening in the transition region of ferritic stainless steel AISI 430 with the 200–240 μm diameter grain size as shown in Fig. 7.2a. The precipitation of carbide throughout the grain boundaries of ferritic phase and weld zone is clearly observed. The carbide developed at the interface of ferritic and weld zone owing to the migration of alloying elements of carbon and chromium from welding electrode with the dilution range of 15–20% [3, 5]. Therefore, a small amount of carbon creates the columnar and coaxial structure in the weldments which are characterized with the presence of austenite and the delta ferrite [4, 5] as shown in Fig. 7.2b. In addition, the development in the transition region between weld and AISI 304L shown in Fig. 7.2c could be referred to as in the previous section of Fig. 7.1a. Now, as already mentioned, the microstructure of similar weldments between 430–430 shows the considerable grain coarsening with the precipitation of carbide at the grain boundary [6, 7] as shown in Fig. 7.2d and f. Both the HAZ regions illustrate the 200–240 μm diameter grain size in transition region, which results in the variation in the hardness values from HAZ to base metal.

Figure 7.3a depicts the micro-hardness results, and there has been a significant variation among the hardness values of the base material, heat-affected zone, and weld zone could be observed for the similar and dissimilar weldments except the 304L-304L one [as in a, b, and c regions of Fig. 7.3a]. The hardness values in similar 430–430 ferritic stainless steel and dissimilar 304L-430 steel weldments increase at the HAZ region compared to the weld zone and base material. This could be attributed to the welding zone with high Cr and Ni contents. Due to the different values of thermal gradient of similar and dissimilar weldments, the occurrence of significant recrystallization in the HAZ region with the uneven distribution of grain coarsening might create the irregular profiles of the hardness.

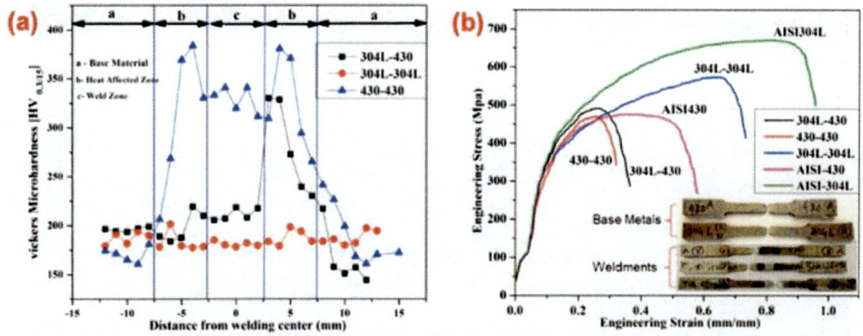

Fig. 7.3 **a** Microhardness of similar AISI 304L-AISI 304L, AISI 430-AISI 430, and dissimilar AISI 304L-AISI 430 shows hardness profile. **b** Stress–strain curve of as-received, dissimilar, and similar weldments between AISI 430 and AISI 304L

Stress–strain diagrams of as-received, similar, and dissimilar weldments of ferritic and austenitic stainless steels are shown in Fig. 7.3b. It shows the three typical stages including elastic deformation, plastic deformation, and fracture phenomenon. The curves also indicate that the UTS and percentage elongation of similar weldments between 304L-304L depict higher load bearing capacity as compared to the similar 430–430 and dissimilar 304L-430 weldments. The weldments of the similar 430–430 and dissimilar 304L-430 specimen indicate that the base metal region adjacent to the weld metal has been stretched preferentially and no significant reduction in cross-sectional area has been found in the welded region. Therefore, the welded specimens show a decrease in the percentage elongation and the fracture happens on base metal side. The observed results might be attributed to the varying Cr and Ni contents and also to the recrystallization behavior of different weldments as mentioned earlier.

7.4 Conclusions

Based on the experimental results the following conclusions have been drawn:

- In the similar 304L-304L weldments, the formation of austenite and delta-ferrite phase with coaxial and columnar structure occurs in the weld region whereas, the development of delta ferrite stringer in the transition region.
- In the dissimilar 304L-430 weldments, the significant grain coarsening in the transition region of ferrite (430) with the 200–240 μm diameter grain size has been observed. While the precipitation of carbide at the interface of weld and the ferritic regions (grain boundaries) can be observed.
- The varying Cr and Ni content and the alteration in the recrystallization behavior ascribe the observed values of toughness, ductility, and hardness for the different

stainless-steel weldments (similar 304L-304L, dissimilar 430-304L, and similar 430–430).

References

1. Madhusudhan Reddy, G., Mohandas, T.: Influence of welding processes on microstructure and mechanical properties of dissimilar austenitic-ferritic stainless-steel welds. Mater. Manuf. Process. **20**, 147–173 (2005)
2. Mirshekari, G.R., Tavakoli, E.: Microstructure and corrosion behavior of multipass gas tungsten arc welded 304L stainless steel. Mater. Des. **55**, 905–911 (2014)
3. Ghorbani, S., Ghasemi, R.: Effect of post weld heat treatment (PWHT) on the microstructure mechanical properties and corrosion resistance of dissimilar stainless steel. Mater. Sci. Eng., A **688**, 470–479 (2017)
4. Ramkumar, K.D., Arivazhagan, N.: Effect of filler materials on the performance of gas tungsten arc welded AISI 304 and Monel 400. Mater. Des. **40**, 70–79 (2012)
5. Satyanarayana, V.V., Madhusudhan Reddy, G.: Dissimilar metal friction welding of austenitic–ferritic stainless steels. J. Mater. Process. Technol. **160**, 128–137 (2005)
6. Guo, G., Shen, Y.: Effect of material position on microstructure and mechanical properties of friction stir welded dissimilar austenite-ferrite stainless steels joints. J. Adhes. Sci. Technol. **35**, 1320–1336 (2020)
7. Sivanantham, A.: Parametric optimization of dissimilar TIG welding of AISI 304L and 430 steel using taguchi analysis. Mater. Sci. Forum **969**, 625–630 (2019)

Chapter 8
Tensile Strength Analysis of 316L Stainless Steel Parts Fabricated via Selective Laser Melting

Meena Pant, Leeladhar Nagdeve, Girija Moona, and Harish Kumar

Abstract Selective laser melting (SLM) is a high-power laser beam process where powder particles are melted to form a compact solid structure. This process is used for the fabrication of complex and lattice structures. The laser power, scanning speed, layer thickness, scanning strategy, and hatch spacing are the most affecting parameters for predicting the component's final quality and mechanical properties. This study examined the impact of process variables on the tensile strength of samples made of 316L stainless steel (SS) using SLM. The microstructure of samples has shown at low-, medium-, and high-energy densities. The Taguchi design and Analysis of Variance (ANOVA) statistical approach has been performed to obtain the optimum parameter for tensile strength. The result showed that the hatch spacing of 100 μm followed by a scanning speed of 900 mm/s and laser power of 350 W are the significant parameters for tensile strength.

Keywords Selective laser melting · 316L stainless steel · Microstructure · Taguchi design · Parameter optimization

8.1 Introduction

Additive technology has progressed and spread quickly worldwide over the past few decades. Selective laser melting (SLM) is the most common metal additive manufacturing process. A solid model is built by melting the powder with a laser beam and following a laser track guided by the CAD geometry. Once the first layer is formed, the next powder layer spreads on the working platform and descends with a piston-cylinder arrangement. The same steps are repeated until the completion of parts [1–4].

M. Pant (✉) · L. Nagdeve · H. Kumar
Mechanical Engineering Department, National Institute of Technology Delhi, New Delhi 110036, India
e-mail: meenapant@nitdelhi.ac.in

G. Moona
CSIR–National Physical Laboratory, New Delhi 110012, India

A. Tewari et al. (eds.), *Proceedings of the 3rd International Conference on Advances in Materials Processing: Challenges and Opportunities*, Springer Proceedings in Physics 293, https://doi.org/10.1007/978-981-99-1971-0_8

It is used in the aerospace, automobile, biomedical, and dental industries to produce complex and customized parts [5, 6]. In SLM, the melt pool's powder properties, process parameters, and thermal history significantly affect the parts' quality and mechanical properties. Wang et al. [7] discussed the thermal gradient effect on the grain structure of 316l SS parts to correlate with mechanical properties. They have estimated that at an energy density of 125 J/mm3, the microhardness is 281.6 HV0.1, elongation is 21.1%, and tensile strength is 590 MPa. Donik et al. [8] investigated the effect of laser power, beam diameter, and scanning speed on the mechanical properties of 316L SS parts. They have reported that improved mechanical properties were obtained at the mid-range of process variables, such as laser power of 200 W, a scan speed of 600 mm/s, and high energy density. Larimian et al. [9] investigated the effect of scanning speed, energy density, and scanning strategy on SLM-fabricated 316l SS samples. They concluded that high scanning speed and single pass with alternate hatch spacing result in good densification. Numerous studies have shown how process factors affect mechanical and physical properties. We have examined the effects of laser power (P), hatch distance (h_d), and scanning speed (V) on the tensile strength of parts made of 316L stainless steel (SS) using the SLM technique. The most important parameter has been predicted using the analysis of variance.

8.2 Material and Methods

8.2.1 Sample Preparation

SLM-based DMP flex 350 metal printers have been used for the fabrication of 316L SS samples. It is a commercial metal printer, where laser power can vary up to 500 W with a focus beam diameter of 60 μm. A fiber-based ytterbium laser has a wavelength of 1070 nm and is used to melt metal and alloys. 316L SS has been used in the poromechanical, biomedical, and dental industries because it has excellent corrosion and mechanical properties. The samples were designed based on the ASTM E8 standard. The design of experiments has been created using Taguchi orthogonal array. It is the best approach to reduce the number of experimental trials through proper tuning between variables, ultimately saving time and less energy consumption for the machine. Three factors and variables have been selected, as shown in Table 8.1 and their combinations in Table 8.2. The energy density was determined using Eq. (8.1), where P refers laser power, V refers scanning speed, h_d denotes hatch distance, and l refers layer thickness.

$$E = (P/(V * h_d * l) \tag{8.1}$$

Table 8.1 Selected factors with ranges

Levels	Factors		
	P (W)	h_d mm	V (mm/s)
−1	250	0.08	700
0	300	0.09	900
1	350	0.10	1100

Table 8.2 Parameters specification for testing

Sample No	V (mm/s)	P (W)	h_d (mm)
S1	700	250	0.08
S2	700	300	0.09
S3	700	350	0.10
S4	900	250	0.09
S5	900	300	0.10
S6	900	350	0.08
S7	1100	250	0.10
S8	1100	300	0.08
S9	1100	350	0.09

8.2.2 *Experimental Setup*

The Tinius Osleen universal testing machine (UTM) has been used to perform the tensile test. The microstructure of samples was analyzed using a Nova Nanosem 450 scanning electron microscope (SEM) with an energy-dispersive spectrometer (EDS). Samples were cut from the gauge length, polished with a different grit of emery paper, and then polished with dry alumina powder solution. The tensile strength of samples has been measured at a strain rate of 2 mm/min.

8.3 Results and Discussion

A typical SLM process creates a solid metallurgical link and permits continuous liquid diffusion based on homogeneous wetting conditions by depositing molten material on a built substrate. After solidification, the sample's temperature is the only variable that affects the material's morphology.

8.3.1 *Microstructure Characterization*

SEM images of samples are shown in Fig. 8.1a–c. The microstructure of samples was observed at high-, medium-, and low-energy densities. Energy-dispersive analysis (EDS) has also been done on samples. The EDS analysis showed all the elements of 316 L stainless steel was present and matched with the standard composition. In Sample 7, unmelted powder and powder sintering have been seen at a low energy density of 45.33 J/mm^3. The balling phenomenon has been seen for sample 3, which was created at a high energy density of 83.33 J/mm^3. Sample 9 was created at 58.33 J/mm^3, and the microstructure is observed to be cellular and columnar.

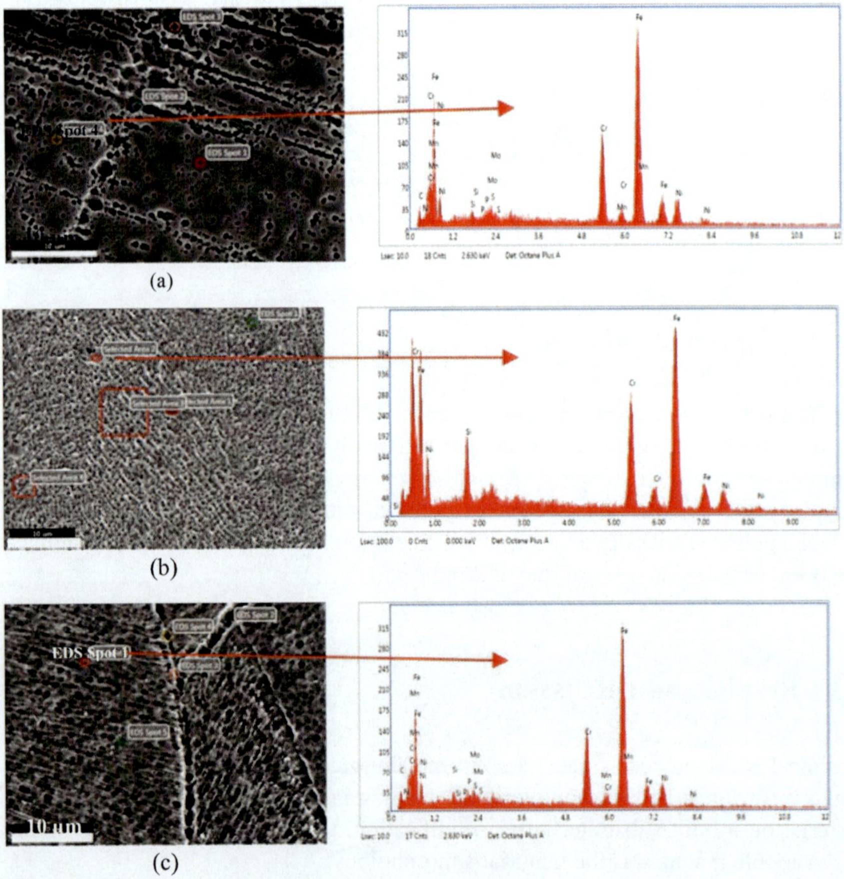

Fig. 8.1 Microstructure of samples **a** Sample 3 at an energy density of 83.33 J/mm3, **b** Sample 7 at an energy density of 45.33 J/mm3, **c** Sample 9 at an energy density of 58.33 J/mm3 (all micrographs at 10 μm resolution)

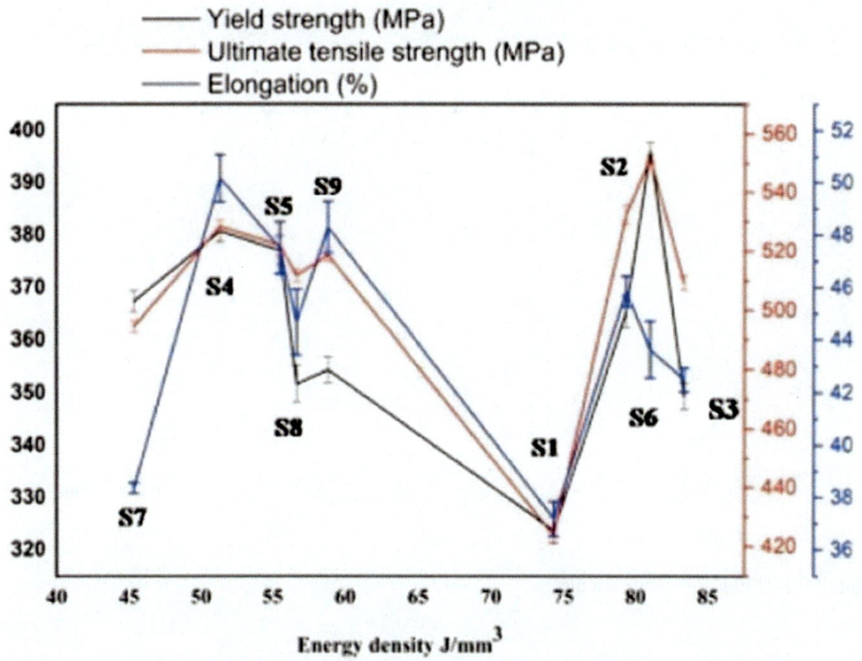

Fig. 8.2 Tensile strength plot against energy density

8.3.2 Tensile Strength Test

The maximum tensile strength is obtained for sample 6, and elongation is obtained for sample 4, as shown in Fig. 8.2. The high tensile strength is obtained at high energy density up to a specific value. The signal-to-noise plot for yield strength "larger is better" was depicted in Fig. 8.3. It demonstrated that a laser of 350 W and a hatch spacing of 0.1 mm offers the best strength for samples at a scanning speed of 900 mm/s. The graph demonstrated that percentage elongation is high at starting energy density, falls for mid-values of energy density, and then begins to increase for high energy densities.

8.3.3 Analysis of Variance (ANOVA)

ANOVA has been done to predict the most significant parameter for tensile strength. The parameters are important if the p-value is less than 0.05 (95% confidence level). Table 8.3 shows that laser power and scanning speed significantly influence tensile strength.

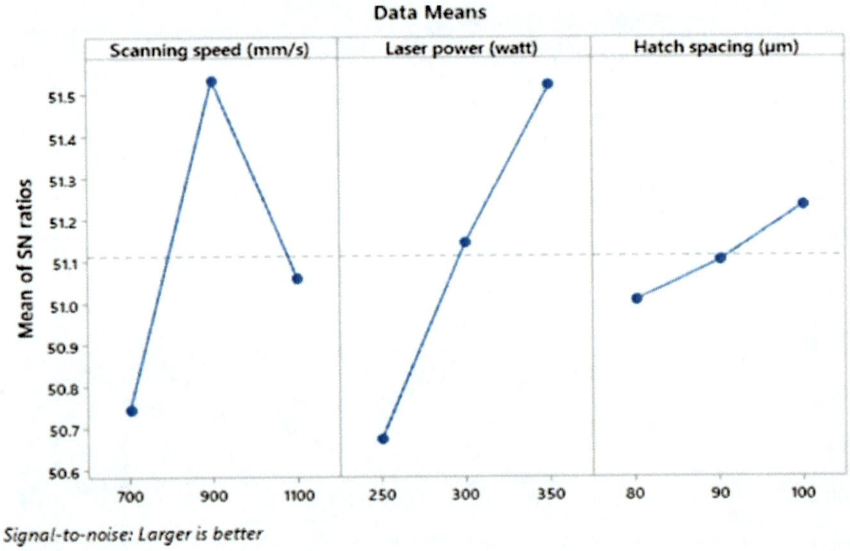

Fig. 8.3 S/N plot for tensile strength

Table 8.3 ANOVA for parameters analysis

Source	DF	Adj SS	Adj MS	F-Value	P-Value
Scanning speed (mm/s)	2	1632.67	816.33	19.75	0.048
Laser power (watt)	2	1842.00	921.00	22.28	0.043
Hatch spacing (μm)	2	108.67	54.33	1.31	0.432
Error	2	82.67	41.33		
Total	8	3666.00			

8.4 Conclusions

The present study focused on tensile strength analysis of 316L SS samples manufactured via the SLM process. The effect of core parameters has been analyzed using ANOVA. The laser power and scanning speed are the most significant parameters. The optimal combination of process parameters, including laser power of 350 W, scanning speed of 900 mm/s, and hatch spacing of 100 μm, respectively.

References

1. Pant, M., Nagdeve, L., Kumar, H., Moona, G.: A contemporary investigation of metal additive manufacturing techniques. Sādhanā **47**(1), 1–19 (2022)
2. Roach, A.M., White, B.C., Garland, A., Jared, B.H., Carroll, J.D., Boyce, B.L.: Size-dependent stochastic tensile properties in additively manufactured 316L stainless steel. Addit. Manuf. **32**, 101090 (2020)
3. Kumar, H., Pant, M., Pidge, P., Nagdeve, L., Moona, G.: Additive manufacturing: The significant role in biomedical and aerospace applications. Indian J. Eng. Mater. Sci. (IJEMS) **28**(4), 330–342 (2021)
4. Pant, M., Nagdeve, L., Moona, G., Kumar, H., & Sharma, A.: Tribological behavior investigation of 316L stainless steel samples processed by selective laser melting. Proc. Instit. Mech. Eng. Part J J. Eng. Tribol. 13506501221101797 (2022)
5. Lodhi, M.J.K., Deen, K.M., Haider, W.: Corrosion behavior of additively manufactured 316L stainless steel in acidic media. Materialia **2**, 111–121 (2018)
6. Liu, Y., Yang, Y., Mai, S., Wang, D., & Song, C.: Investigation into spatter behavior during selective laser melting of AISI 316L stainless steel powder. Mater. Des. **87**, 797–806 (2015)
7. Wang, D., Song, C., Yang, Y., Bai, Y.: Investigation of crystal growth mechanism during selective laser melting and mechanical property characterization of 316L stainless steel parts. Mater. Des. **100**, 291–299 (2016)
8. Donik, Č, Kraner, J., Paulin, I., Godec, M.: Influence of the energy density for selective laser melting on stainless steel's microstructure and mechanical properties. Metals **10**(7), 919 (2020)
9. Larimian, T., Kannan, M., Grzesiak, D., AlMangour, B., Borkar, T.: Effect of energy density and scanning strategy on densification, microstructure and mechanical properties of 316L stainless steel processed via selective laser melting. Mater. Sci. Eng. A **770**, 138455 (2020)

Chapter 9
Investigating the Effects of Induction Heating on Friction Stir Welding of Low-Carbon Steel

Pankaj Kaushik, Rajdev Singh, Arun Jena, and Dheerendra Kumar Dwivedi

Abstract Friction stir welding was primarily developed for soft and low-temperature materials. With the help of emergent technologies and growing demand, this joining process can also be used for high-temperature materials like copper, steel, and titanium. Conventional friction stir welding of steel demands high-temperature stirring where the tool has to withstand high wear, which reduces the tool's life. Reducing the plate hardness and increasing its flowability before interaction with the tool can prove beneficial for improvement in tool life. The present study establishes an induction-assisted preheating setup just ahead of the tool. Friction stir welding of steel with and without induction preheating is performed, and the effects of preheating on joint properties are investigated. The welded joints are compared depending on their weld thermal cycles, microstructural characterization, and mechanical properties.

Keywords Steel · FSW · Induction · Preheat

9.1 Introduction

The friction stir welding (FSW) process is a solid-state joining process mainly invented to join low melting point materials such as Al, Mg, and Cu alloys. But with time, the process has evolved, and the feasibility of the same welding process for high melting temperature metals like steel has been investigated. Steel fusion welding is associated with various complications like porosity, solidification cracking, solute redistribution, etc. These problems mainly arise during the solidification of the molten metal in the fusion welding process. Unlike fusion welding, FSW is advantageous as the welding occurs only due to thermomechanical action, which happens below the material's melting temperature [1, 2]. During FSW of soft metals, the welding takes place at a lower temperature, so there are no issues with the tool's wear and life

P. Kaushik (✉) · R. Singh · A. Jena · D. K. Dwivedi
Department of Mechanical and Industrial Engineering, Indian Institute of Technology Roorkee, Uttarakhand 247667, India
e-mail: pkaushik4@me.iitr.ac.in

A. Tewari et al. (eds.), *Proceedings of the 3rd International Conference on Advances in Materials Processing: Challenges and Opportunities*, Springer Proceedings in Physics 293, https://doi.org/10.1007/978-981-99-1971-0_9

[3]. Along with using a high-hot hardness tool, bringing the flow stress of the hard material to a lower level might prove beneficial for tool life [4]. Preheating during FSW of hard materials positively affects the welding process, heat generation, material flow, weld mechanical properties, and morphology of the welded surface. The auxiliary energy source can be used during FSW to preheat the plates. Preheating during FSW can be realized via electrical, laser, induction, or arc assistance [5]. The material flow increased and ultimately reduced the tendency of void formation in preheated FSW due to improved stirring action [6].

The research literature available on solid-state welding of steel plates is scarce. Developing sound weld joints of different steel grades is challenging because of tool wear and tool life. These problems can be eradicated by preheating the work material ahead of the tool. Preheating steel plates before welding by arc and laser methods may contaminate the weld and prove costly. Further, the measurement of the thermal cycle and its effects on cooling rate in steel welding was not explored in the literature. So, in this work, friction stir welding of low-carbon steel plates was carried out with and without preheating. The current study used a non-contact induction heater to preheat the plate to attain cost-effective and sound weld joints. The mechanical properties of the joints formed with and without preheating were investigated. It was found that the properties were significantly affected due to coarsening of the grains in the case of preheating. Weld thermal cycles were captured during welding, and the cooling rate was compared in both cases. Tensile, hardness, and optical microscopy have been used to correlate the changes in mechanical properties due to preheating.

9.2 Materials and Experimental Procedure

The steel plates of size 150 mm × 50 mm × 3 mm were joined by the FSW process in abutting condition. The chemical composition of the steel plate is given in Table 9.1.

The tungsten carbide (WC) tool was used for the welding having a tapered cylindrical pin of diameter 8–10 mm with a pin length of 2.72 mm. The parameters used during welding were 508 rpm, 26 mm/minute traverse speed, and a tilt angle of 2 degrees. A non-contact type induction heater of 15 kW and a frequency of 30–100 kHz is employed to preheat the steel plates. The tool profile and schematic of the preheating setup are shown in Fig. 9.1.

Table 9.1 Chemical composition (wt. %) of steel plate

Elements	Fe	C	Mn	S	Si	P	Cr	Ni
wt. %	Balance	0.242	1.05	0.015	0.096	0.019	0.046	0.023

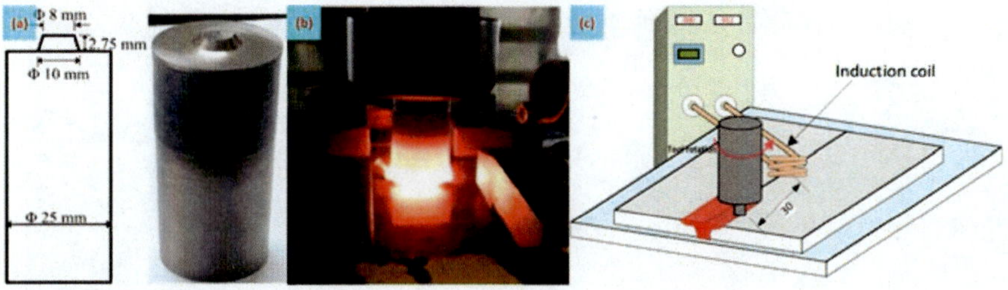

Fig. 9.1 **a** Tool profile, **b** Tool heating during welding, and **c** Schematic of FSW with induction heating on steel plate

9.3 Results and Discussion

The actual welded plates and transverse section of the welds for normal and preheating cases are shown in Fig. 9.2. Transverse cross sections of the welds were analyzed by stereoscopy, through which the intermixing zone was identified. Three distinct zones, namely base metal, thermomechanically affected zone (TMAZ), and stir zone, were observed during the microstructural investigation of friction stir weld joints. The optical micrograph of base metal, i.e., structural steel, contains polygonal ferrite and a small amount of fine pearlite identified by dark points. Due to preheating and plastic deformation during friction stir welding, the temperature exceeds the limit of lower critical temperature (727 °C), confirmed by the thermocouple readings. The temperature attained was sufficient to result in the recrystallization of grain and grain growth. The TMAZ of the welded joint without preheating contains coarse-grain ferrite and pearlite, which are shown in Fig. 9.2. With induction preheating at a preheating current of 300 A, the grains became slightly coarser because of the lower cooling rate. The average grain size of equiaxed ferrite in base metal was 23.44 μm.

The stir zone contains grain boundary ferrite and ferrite–carbide aggregate along with ferrite aligned with the second phase. In the stir zone of the welds, the average grain size reduced marginally because of dynamic recrystallization during friction stir welding. The average grain size of the ferrite of the stir zone developed without preheating was 16.41 μm; however, when a preheating current of 300 A was used, having all other parameters unchanged, the grain size increased to 18.65 μm. This increment in the average grain size in the stir zone was due to the slow cooling rate.

A K-type thermocouple is placed approximately 1.5mm from the edge of the stir zone and at the middle of the length of the weld. The cooling rate before the temperature reached 300 °C was calculated according to the formula: Cooling rate $= (T_e - 300)/\Delta t$, Where T_e is the maximum temperature, and Δt is the time when the temperature decreased to 300 °C from the maximum temperature. At the point of interest, the normal weld showed a cooling rate of 15.27 °C/sec, but after preheating, the cooling rate of the same point decreased to 10.106 °C/sec. The peak temperature, the time taken to reach 300 °C, and the cooling rate are listed in Table 9.2.

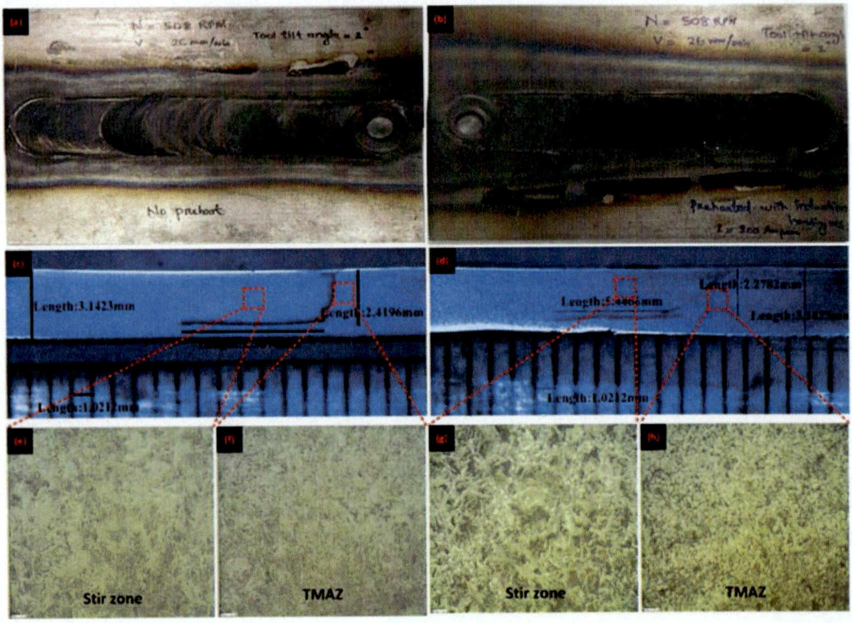

Fig. 9.2 Weld image of steel FSW in **a** normal condition, **b** preheat condition, **c**, **d** Stereo images of weld cross section, and **e**, **f**, **g**, **h** Optical micrographs of stir zone and TMAZ

Table 9.2 Cooling rate calculation with and without preheated FSWs

Weld joint	Peak temperature (°C)	Time from peak temperature to 300 °C (sec)	Cooling rate (°C/sec)	Remarks
Without preheat	911	40	15.275	Cooling rate decreases with preheating
With preheat (Preheating current I = 300A)	1058	75	10.106	

The weld thermal cycles of normal and induction-assisted FSW joints of steel are shown in Fig. 9.3a. The induction-assisted weld thermal cycle showed a double-humped curve, where the first hump of 900 degrees attained from the preheater effect.

The tensile test was performed, and it was observed that the fracture location of almost all samples was outside the gauge length, i.e., from the base metal. The engineering stress–strain curves for normal weld and preheated weld are shown in Fig. 9.3b. The zones consist of varying grain sizes and phase structures, resulting in different tensile properties. The yield strength of base metal was 248 MPa. The ultimate tensile strength (UTS) of all the weld samples was higher than the base

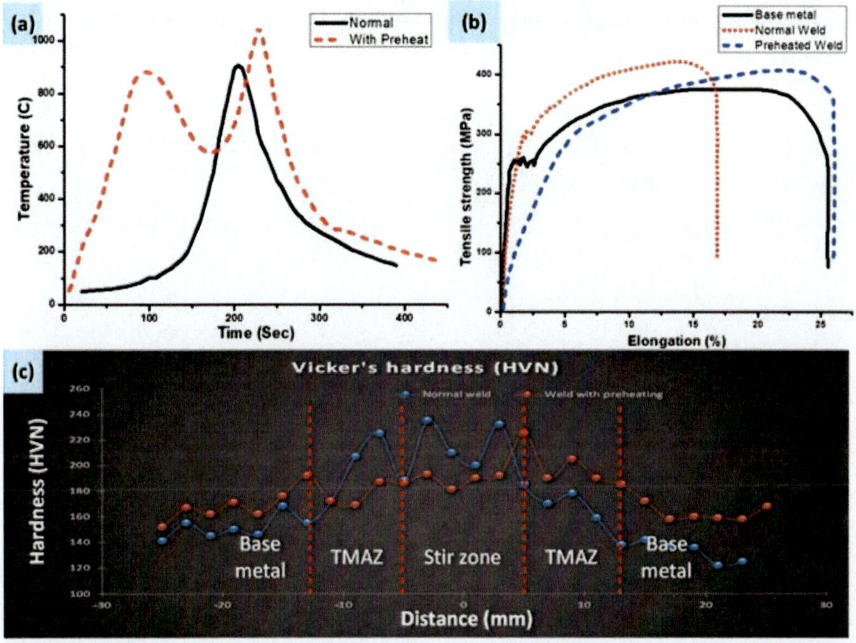

Fig. 9.3 **a** Weld thermal cycle, **b** Tensile test curves, **c** Hardness profile of normal and preheated FSW joints

metal, and also the percentage elongation of preheated sample was comparable to the base metal.

The microhardness of normal friction stir welded joints and preheating was measured. The effect of preheating during FSW on hardness distribution is shown in Fig. 9.3c. In the induction-assisted preheated weld joint, the hardness curve came out to be gradually varying from TMAZ to stir zone to TMAZ again. In contrast, normal weld showed a steep rise in stir zone hardness. The hardness reduction due to preheating could be correlated to the formation of new coarse and recrystallized grains.

9.4 Conclusions

The present work describes the weld joint developed by the FSW process with and without induction preheating. The following observations can be considered significant based on the work accomplished in the present study:

- On the application of the induction heater ahead of FSW, preheating of the steel plates take place, which reduces the material's yield strength and, in turn, allows easy stirring of the hard steel metal.

- A decrement in the cooling rate of the preheated weld was observed, and the average grain size of the stir zone increased by grain growth due to the slow cooling rate.
- The mechanical properties (improved tensile strength and a gradual hardness variation) were enhanced due to induction preheating in steel FSW. The tool wear was also reduced using preheating in FSW.

References

1. Kaushik, P., Dwivedi, D.K.: Influence of hook geometry on failure mechanism of Al6061-galvanized steel dissimilar FSW lap joint. Arch. Civ. Mech. Eng. **22**, 149 (2022). https://doi.org/10.1007/s43452-022-00471-z
2. Kaushik, P., Dwivedi, D.K.: Heat generation and steel fragment effects on friction stir welding of aluminum alloy with steel. Mater. Manuf. Proc. **00**(00), 1–13 (2023). https://doi.org/10.1080/10426914.2023.2187827
3. Kaushik, P., Dwivedi, D.K.: Materials today : proceedings Al-steel dissimilar joining : challenges and opportunities. Mater. Today Proc. **62**, 6884–6899 (2022). https://doi.org/10.1016/j.matpr.2022.05.211
4. Kaushik, P., Kumar Dwivedi, D.: Induction preheating in FSW of Al-Steel combination. Mater. Today Proc. **46**, 1091–1095 (2021). https://doi.org/10.1016/J.MATPR.2021.01.438
5. Padhy, G.K., Wu, C.S., Gao, S.: Auxiliary energy assisted friction stir welding–status review. Sci. Technol. Weld. Join. **20**, 631–649 (2015). https://doi.org/10.1179/1362171815Y.0000000048
6. Yaduwanshi, D.K., Bag, S., Pal, S.: Effect of preheating in hybrid friction stir welding of aluminum alloy. J. Mater. Eng. Perform. **23**, 3794–3803 (2014). https://doi.org/10.1007/s11665-014-1170-x

Chapter 10
Microstructural Evolution During Homogenization of Ni–Cu–Al–Ti Alloy

Anand Udare, S. Chenna Krishna, Pravin Muneshwar, and Bhanu Pant

Abstract In the present investigation, the macrostructure and microstructure of the as-cast and homogenized Ni–Cu–Ti–Al were examined using optical microscopy and scanning electron microscopy equipped with energy-dispersive X-ray spectroscopy. The macrostructure showed a classical solidification pattern with three distinct zones: chilled equiaxed at the mould/ingot interface, columnar at the periphery and equiaxed at the centre. The as-cast structure was characterized by the dendritic structure and interdendritic regions was rich in copper. In addition, primary TiC particles were primarily present in interdendritic regions. The homogenization cycle was optimized based on the microstructural examination. The hot working temperature was selected based on the hot compression of cylindrical specimens in a hydraulic press. Based on these studies, a homogenized ingot of 80 mm dia. × 100 mm was hot worked in the temperature range of 1100–1000 °C to produce rods of 30 mm diameter. Heat treatment studies were conducted to optimize solution treatment and the aging cycle. The typical mechanical properties in the aged condition were UTS—1025 MPa, YS—758 MPa, % Elongation—23% and hardness—270 HV.

Keywords Monel K500 · Homogenization · Hot deformation · Microstructure

10.1 Introduction

Monel K500 is a Ni–Cu–Ti–Al alloy discovered by the International Nickel Company [INCO]. Ni–Cu shows a completely soluble structure as in isomorphous as Ni and

A. Udare (✉)
College of Engineering, Pune, India
e-mail: udaream20.meta@coep.ac.in

S. Chenna Krishna · P. Muneshwar
Materials Processing Division, Vikram Sarabhai Space Centre, Trivandrum, India

B. Pant
Kalyani Centre for Technological Innovation, Pune, India

A. Tewari et al. (eds.), *Proceedings of the 3rd International Conference on Advances in Materials Processing: Challenges and Opportunities*, Springer Proceedings in Physics 293, https://doi.org/10.1007/978-981-99-1971-0_10

Cu in binary solution follow Hume-Rothery rules. Apart from two primary elements, the alloy has an addition of titanium and aluminium which by age hardening mechanism forms sub-microscopic precipitates of Ni_3 (Ti, Al) throughout the matrix thus enhancing alloy properties [1, 2]. Due to a combination of strength and corrosion resistance, they are used for products like chains and cables, fasteners and springs for naval duty and pump and valve components for chemical processing [2]. Additionally, due to their high ignition resistance and ability to maintain their toughness at cryogenic temperatures, these alloys make excellent candidates for structural machinery parts and the storage of liquefied gases in the aerospace sector [3].

Earlier studies on Monel K500 were focused on aging characteristics [4, 5], hot deformation [6, 7] general corrosion [8] and tribocorrosion [9, 10]. Very limited data is available in the open literature on ingot metallurgy and thermomechanical processing. An attempt has been made for:

- Vacuum induction melting of Monel K500;
- Establishing appropriate homogenization and thermomechanical processing parameters;
- Correlation between the processing parameters, microstructure and properties.

10.2 Material and Methodology

A Ni–Cu–Ti–Al (Monel K500) was melted (24 kg) in a magnesia crucible using vacuum induction melting route. The raw materials consisting of pure metals such as (Ni, Cu, Al and Ti) were weighed to achieve aimed composition. Sequential addition of charge was employed to avoid loss of reactive elements. Ni and Cu were stacked systematically in the magnesia crucible to ensure maximum coupling. Melting of the charge was done by progressively increasing power. Sufficient time was provided for melting of charge and degassing. After ensuring the complete melting of the charge, Ti and Al are added through an alloy charger. The molten melt was poured into a split-type mild steel mould coated with magnesia powder. After providing sufficient time for solidification, the top and bottom of the cast ingot of size 80 mm dia. × 100 mm were cut to remove the piping and segregation.

Specimens were drawn from the top and bottom of the ingot for chemical analysis. The chemistry of the alloy was analysed using Hitachi master Pro2 Optical Emission Spectroscopy which is listed in Table 10.1. A section of 20 mm thickness was cut from the top of the cast ingot for evaluation of the macrostructure. The specimen was ground and etched in the solution of HCl and HNO_3, known as aquaregia in a ratio of 1:3 by holding for 10–15 min. A macro-etched specimen was neutralized using NaOH solution.

Homogenization studies were performed in the temperature range of 1100 °C, 1150 °C and 1200 °C and the holding time: 5 h. Hot deformation studies were performed on as-cast and homogenized specimens using specimens of 8 mm dia. × 12 mm height. All the specimens were soaked at a temperature of 1000 °C, 1050 °C and 1100 °C for a holding time of 30 min. All specimens were subjected to a single

Table 10.1 Chemical composition of the as-cast Monel K 500 alloy

Cu	Al	Ti	Fe	C	Ni
31.33	2.85	0.55	0.88	0.22	Bal

upsetting operation for the true strain of about 0.3 at a pressing speed of 10 mms^{-1} and water quenched. Specimens for microstructural examination were prepared by polishing using emery paper, alumina paste and 1 μm diamond paste. Glyceregia in the ratio (HCl: HNO_3: glycerol = 3:1:2) was used as an etchant. The microstructure was investigated using an inverted optical microscope, Metscope Pro. The SEM analysis was done using Tescan Vega 3LWH equipped with EDS.

Further after the determination of suitable working conditions, the ingot was converted to 30 mm diameter rods following multi-step drawing operations. Solution treatment and aging studies were performed using hardness measurement. Upon determining an optimum solution treatment and aging cycle, samples were cut for tensile testing with gauge length in the direction parallel to the rolling direction. Tests were performed on UTM following ASTM E8.

10.3 Results and Discussion

10.3.1 Macrostructure and Microstructure

Figure 10.1 shows the macrostructure at the top of the cast ingot and the microstructure at the centre and periphery in two directions. Figure 10.1a demonstrated a very fine equiaxed chilled zone cell structure at the periphery which was in contact with mould walls due to the highest cooling rate. The grain structure changed to a columnar type as the solidification plane progressed inside along a radial direction. Moving further near to centre feathery equiaxed grains were formed. Such change in grain structure is mainly attributed to variation in the G/R ratio (thermal gradient/ growth rate). It can be concluded a non-planar solidification occurred. Further, the L/D ratio of the grains was measured perpendicular to the mould wall. It had a mean value of 3.99 and 1.472 for columnar and equiaxed grains, respectively. The outer chilled zone extended for 1–2 cm, while the central equiaxed grains extended for a region of size 3–4 cm approximately.

The enlarged view of the microstructure present along the periphery and centre is along longitudinal and transverse directions shown in Fig. 10.1b–e. From the micrographs, it can be concluded that the severity of the segregation is quite less along both directions at the periphery. Due to solute rejection, most of the solute is concentrated in the central region. The columnar to equiaxed transition (CET) in the longitudinal direction is evident in Fig. 10.1c. The dendritic feature is eliminated along the transverse direction, as seen in Fig. 10.1e. The cell sizes for the centre and the periphery were 115.4 and 127.6 μm, respectively.

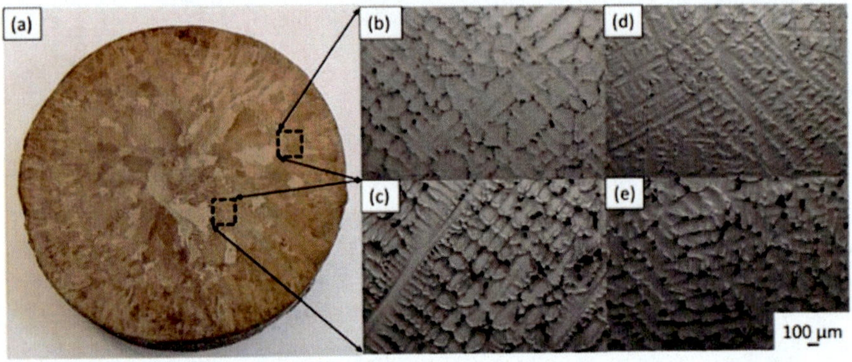

Fig. 10.1 Macrostructure and microstructure of the cast ingot: **a** Macrostructure at cast ingot (top); Microstructure **b** periphery-L **c** centre-L **d** periphery-T, and **e** centre-T; L-longitudinal and T-transverse

Additionally, it can be seen (Fig. 10.1d) that the major dendrites in the peripheral part of the ingot were more closely spaced compared to the centre. The reported dendrites had shorter overall primary lengths because they appeared to stop growing at frequent intervals. This could be due to the occurrence of more nucleation before the dendrite tip advancement. The secondary dendritic arm spacing (SDAS) at the periphery and centre was 76 μm and 57 μm, respectively.

The SEM image shows the dendritic structure and primary particle along with the line scan (Fig. 10.2). The line scan shows compositional variation across the dendrites (Fig. 10.2a). There is variation in the Ni and Cu across the line, especially at the locations indicated by dashed lines (1,2,3). The white regions corresponding to dendrites showed enrichment of Cu. SEM micrograph and corresponding EDS spectra of the primary particles are shown in Fig. 10.2b. From the EDS spectra and quantitative analysis, it can be ascertained the primary particles (Spot 1) are TiC carbides and Spot-2 confirms to matrix composition of Ni–32Cu–3Al (wt.%). Thus, it can be concluded that particles at interdendritic regions are TiC carbides.

10.3.2 Homogenization and Hot Working

The optical micrographs of specimens homogenized and hot worked at different temperatures are shown in Fig. 10.3. As evident from Fig. 10.3a, the extent of segregation gradually reduced with the increase in the homogenization temperature (1100, 1150 and 1200°C). It is noteworthy to mention that the centre showed higher segregation compared to the periphery and the same pattern was observed after homogenization. With the increase in temperature, the severity of segregation reduced but at the cost of grain coarsening. At 1200 °C, there was significant grain growth and severe oxidation. On the other hand, at 1100 °C, there was a reduction in segregation

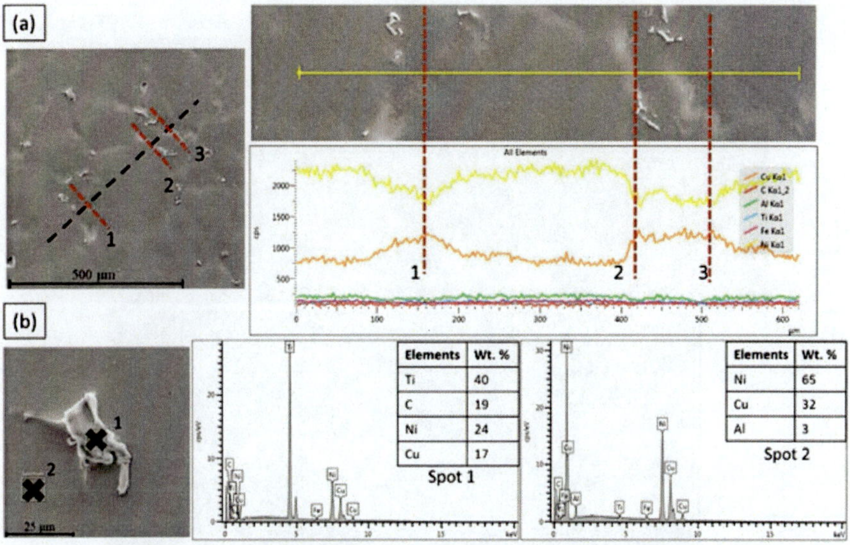

Fig. 10.2 **a** SEM line scan showing microsegregation in the as-cast sample **b** Typical EDS qualitative analysis spectrum of phases like Ti–C particle, Ni–Cu matrix phase (numbers indicating analysed spots)

but the remnant dendritic structure was evident. Hence, 1150 °C was selected as the optimum temperature for homogenization.

The microstructure of the as-cast and homogenized specimen subjected to hot deformation in the temperature range of 1000–1100 °C is shown in Fig. 10.3b. The as-cast after hot working showed remnant cast structure and grain boundary cracking was observed at a temperature of 1000 °C and microporosity at 1100 °C. All the deformed specimens showed the characteristic feature of the deformation bands created followed by texture-like serrations. The homogenized samples did not show visible cracks on the surface and the severity of the localized stress zone was reduced. Based on these observations, the homogenized ingot was hot working in the temperature range of 1100–1000 °C and converted to a 30 mm diameter rod through multiple steps of upsetting and drawing. The microstructure of the hot worked rod with equiaxed grains is shown in Fig. 10.3c.

10.3.3 Hardness and Strength

The influence of solution treatment and aging on the hardness of the Monel K500 is listed in Table 10.2. Solution treatment was performed in the temperature range of 750–950 °C. In the as-forged condition, hardness was 234 HV. With an increase in the solution treatment temperature, a reduction in hardness was observed from

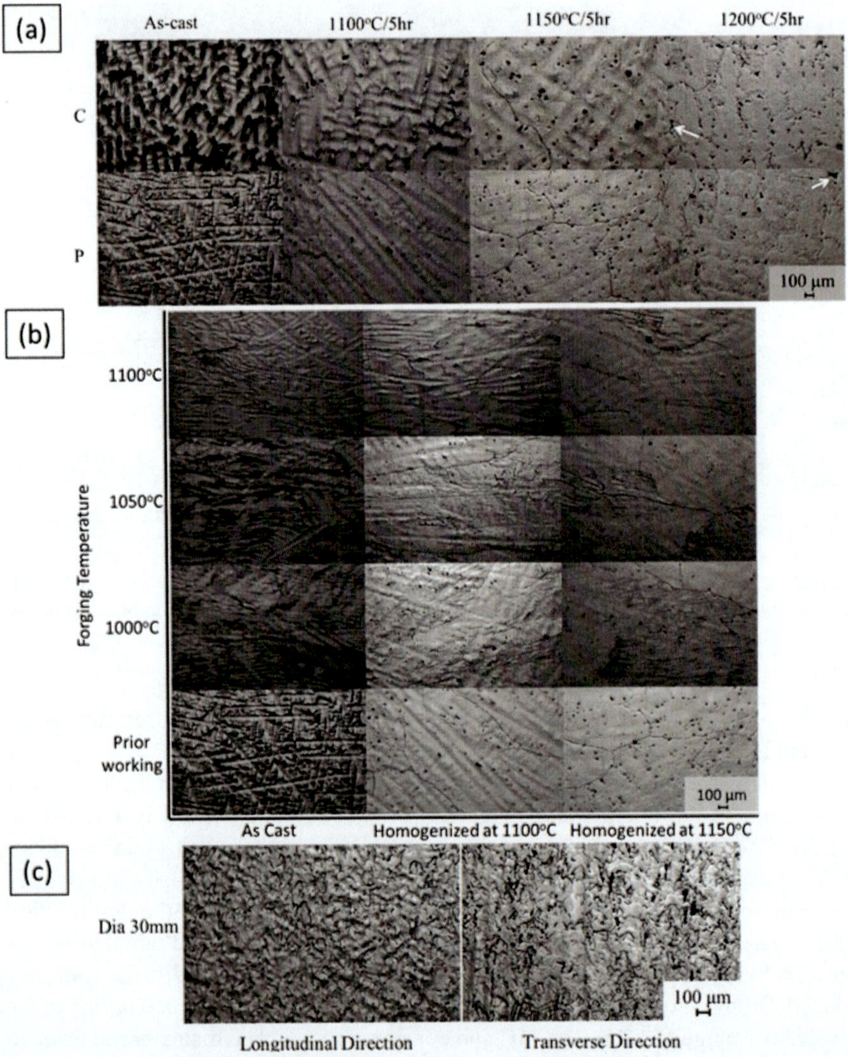

Fig. 10.3 **a** Optical Micrographs of as-cast and homogenized specimens (C—centre and P—periphery), **b** Microstructural evolution after hot working at different temperatures, **c** Optical micrographs in the hot forged rods (30 mm dia.)

199 HV at 750 °C to 134 HV at 950 °C. The decrease in hardness may be attributed to the reduction in the stresses induced due to hot working and grain coarsening. A similar trend was observed after the aging of solution-treated specimens. As-forged and direct-aged specimens showed a hardness of 304 HV. The hardness of the ST750 and the aged specimen was 290 HV, decreasing to 197 HV for the ST950 specimen. The microstructure of the hot worked rod shown in Fig. 10.3c depicted equiaxed

Table 10.2 Influence of solution temperature and aging on the hardness of Monel K500

Condition	ST hardness (HV)	Aged at 600 °C/16 h + furnace cool to 482 °C by cooling rate of 11 °C/hr (HV)
As forged	234 ± 2.50	304 ± 2.23
750 °C/30 min/WQ (ST750)	199 ± 2.70	290 ± 2.51
800 °C/30 min/WQ (ST800)	164 ± 1.10	258 ± 5.04
850 °C/30 min/WQ (ST850)	149 ± 0.58	234 ± 2.85
900 °C/30 min/WQ (ST900)	140 ± 1.64	208 ± 3.35
950 °C/30 min/WQ (ST950)	134 ± 0.38	197 ± 1.82

austenitic grains with an average grain size of 48 μm (ASTM No. 6). The tensile properties of the 30 mm diameter rods were evaluated in optimum aged condition on the round tensile specimen as follows: ultimate tensile strength of 1025 MPa, yield strength of 758 MPa and % elongation of 23%.

10.4 Conclusions

- Macrostructure showed a classical solidification pattern with three distinct zones: chilled equiaxed, columnar and equiaxed.
- As-cast microstructure had dendritic features. The mean secondary dendritic arm spacing at the periphery and centre was 76 μm and 57 μm, respectively.
- SEM–EDS analysis confirmed that dendrites were rich in copper and TiC particles were present at the interdendritic regions
- Hardness decreased with the increase in the solution treatment temperature. After aging, there is a significant increase in hardness with a maximum value of 304 HV.
- Microstructure of the hot-worked rod showed equiaxed austenitic grains with an average grain size of 48 μm.
- Tensile properties for the final processed condition were UTS, YS and elongation in the aged condition were reported to be 1025 MPa, 758 MPa and 23%, respectively.

Acknowledgements The author gratefully acknowledges his dissertation guides' support, guidance, and encouragement. The author would like to thank the staff at MPD/VSSC for the support rendered in the project.

References

1. Davis, J.R.: ASM speciality handbook: nickel, cobalt, and their alloys **38**(11) (2001)
2. Special Metals Corporation, "Monel Alloy K-500". Spec. Met. 1–20 (2004)
3. Kostryzhev, A.: Nickel-copper alloys
4. Dey, G.K., Mukhopadhyay, P.: Precipitation in the NiCu-base alloy Monel K-500. Mater. Sci. Eng. **84**(C), 177–189 (1986)
5. Dey, G.K., Tewari, R., Rao, P., Wadekar, S.L., Mukhopadhyay, P.: Precipitation hardening in nickel-copper base alloy Monel K 500. Metall. Trans. **24**, 2709–2719 (1993)
6. Zhang, X., Chen, W.: Experimental research on the heating performance of Monel k-500 alloy. Adv Mat.Res. **572**, 273–277 (2012)
7. Zhang, X.P., Chen, W.Z.: Experimental research on the hot working property of Monel K-500 alloy. Adv. Mat. Res. **712–715**, 87–93 (2013)
8. Jun, C., Fengyuan, Y.: Corrosive wear performance of monel k500 alloy in artificial seawater. Tribol. Trans. **56**(5), 848–856 (2013)
9. Krawczyk, J., Frocisz, L., Matusiewicz, P., Madej, M.: The effect of the microstructure on the tribological properties of the monel K-500 alloy. In: Metal 2015—24th ICMM, pp. 1326–1334 (2015)
10. Chen, J., Yan, F.Y., Chen, B.B., Wang, J.Z.: Assessing the tribocorrosion performance of Ti-6Al-4V, 316 stainless steel and Monel K500 alloys in artificial seawater. Mater. Corros. **64**(5), 394–401 (2013)

Chapter 11
Effect of Nickel Addition on the Evolution of Microstructure of ZA-27 Alloy

K. Anusha Raj, G. Ramesh, and Babu Rao Jinugu

Abstract Aeronautical and automobile industries demand to develop engineering materials with high strength-to-weight ratio due to fuel economy. Modification of microstructure through alloying in order to alter mechanical properties is an effective method in metallurgical field. In the present work, the effect of nickel addition (1 wt%) and cooling rate during solidification (2.5 and 10 °C/min) on microstructure evolution and hardness of ZA-27 alloy were investigated. Increase in cooling rate during solidification of ZA-27 alloy resulted in a decrease in size and change in morphology of the primary phase. The addition of nickel resulted in rosette-type primary α phase, less amount of eutectoid phase, and formation of metallic compound (Al_3Ni_5). Cooling rate does not have any influence on the refinement of microstructure of ZA-27 alloy in the presence of nickel. However, it changes the shape of metallic compounds. Increase in cooling rate and addition of nickel resulted in an increase in the hardness value of ZA-27 alloy.

Keywords ZA-27 · Nickel · Microstructure · Solidification · Cooling rate

11.1 Introduction

Zinc base cast alloys, commonly referred to as "ZA" alloys, are widely used in technological applications due to their superior mechanical, physical, and tribological properties, corrosion resistance, high damping capacity, castability, environmentally friendliness, and low cost–energy ratio. ZA-27 alloy is the subgroup

K. A. Raj · G. Ramesh (✉)
Department of Metallurgical and Materials Engineering, Rajiv Gandhi University of Knowledge Technologies, RK Valley Campus, Kadapa 516330, India
e-mail: gramesh.rmg@gmail.com

B. R. Jinugu (✉)
Department of Metallurgical Engineering, AU College of Engineering-Andhra University, Visakhapatnam 530003, India
e-mail: brjinugu@gmail.com

A. Tewari et al. (eds.), *Proceedings of the 3rd International Conference on Advances in Materials Processing: Challenges and Opportunities*, Springer Proceedings in Physics 293, https://doi.org/10.1007/978-981-99-1971-0_11

of ZA alloys which contains nearly 27 wt% of aluminum. Generally, ZA-27 alloy components are produced by pressure die casting and/or gravity die casting. ZA-27 alloy provides excellent strength with low density among all other ZA alloys and is used as alternative material to bronze in bearing and bushing applications [1, 2]. The alloying/ceramic particle addition has a significant influence on microstructural modifications and thus improvement in mechanical properties of the components. Krupiński [3] investigated the effect of cerium addition on the microstructure of Zn–Al–Cu alloys and reported that eutectic size reduction and morphology change from dendritic to "tweed" of α' phase precipitates by the addition of a small amount of cerium (0.68 wt%). Dixin Yang et al. [4] investigated the microstructure and wear behaviour of ZA-30 with lanthanum. Addition of 0.3% lanthanum resulted in metallic compounds and also improved abrasive wear resistance. Ramesh et al. [5] studied the cooling behavior and microstructure of chill cast ZA-8 modified with 0.5 wt% manganese. Thermal analysis showed that the manganese addition resulted in increased liquidus temperature, decreased eutectic and eutectoid temperatures, decreased heat of solidification, and lowered wettability of ZA-8 alloy. The manganese-modified ZA-8 alloy showed higher fraction of dendrite phase and lower amount eutectic phase compared to the unmodified ZA-8 alloy. Literature clearly indicates the addition of alloying elements and cooling rate during solidification have a significant effect on the microstructure and mechanical properties of zinc–aluminum alloy casting. Though ZA-27 alloys exhibit easy castability, excellent strength, and corrosion, the dimension stability and mechanical strength of ZA-27 alloys at higher temperature were poor. Hence, the addition of high melting alloying elements and/or ceramic particles to ZA-27 alloy is necessary to improve high-temperature creep strength and dimension stability. The present work is aimed to study the effect of nickel additions and cooling rate during solidification on the microstructure and hardness of ZA-27 alloy.

11.2 Materials and Methods

Commercial pure zinc, aluminum, and Al–10% Ni master alloy were procured from M/s MatRICS, Tamilnadu, for preparing the alloy. The present investigation aimed to modify ZA-27 alloy with 1% nickel. Accordingly, calculated weight percentages of pure zinc, pure aluminum, and Al–10% Ni alloy were measured using an electronic weighing balance. The known amount of aluminum and Al–10% Ni alloy were taken into graphite crucible and then melted in an electrical resistance furnace at 750 °C. On melting of aluminum and Al–10% Ni alloy, the preheated zinc ingot piece (200 °C in a muffle furnace) was added to melt and temperature of the furnace was set to 600 °C for 30 min. Once the alloy was melted to liquid, the melt was stirred using a graphite rod, poured into clay crucibles (capacity 150 g) and then allowed to cool down to room temperature. The chemical composition of the prepared alloy was determined using Inductive coupled plasma optical emission spectrometry (ICP-OES) (model GBC 932 plus). Table 11.1 shows the chemical analysis of the prepared alloy. The

Table 11.1 Chemical analysis of prepared alloy

Element	Mg	Si	Al	Fe	Ni	Zn
Zn–27Al	0.19	0.06	26.91	0.08	0.00	Balance
Zn–27Al–1Ni	0.10	0.19	26.56	0.07	1.06	Balance

solidified alloys were then once again re-melted to 600 °C in a programmable muffle furnace, held at that temperature for 45 min, and then cooled to room temperature at different cooling rates of 2.5 and 10 °C/min. The alloys solidified at different cooling rates were sectioned using low-speed cutting machine for microstructural study. The sectioned samples were polished using 180#, 220#, 400#, 600#, 1000#, and 1200# silica carbide emery papers and finally with disc polishing using 3 μm granulation diamond paste to obtain mirror finish. The polished samples were etched using a solution (Water—100 mL, Cr_2O_3—20 g, and Na_2SO_4—5 g) and then microstructures of the samples were analyzed using Leica DM5000 Metallurgical Microscope. X-ray Diffraction (XRD) analysis was carried out using Bruker D8 Advanced machine to understand the phase formation. The diffraction study was conducted at a scan speed of 2°/min in a step size of 0.02. The hardness of the castings was measured using a Vickers hardness tester. A load of 5 kg was applied for a dwell time of 15 s using a diamond indenter.

11.3 Results and Discussion

Figure 11.1 shows the microstructure of ZA-27 alloy solidified at different cooling rates of 2.5 °C/min and 10°C/min. Microstructure consists of the heterogeneous structure due to complex solidification of ZA-27 alloy via $L \rightarrow \alpha + L \rightarrow \beta \rightarrow \alpha + \beta \rightarrow \alpha + \eta$. The coarse primary α phase was found to be present in eutectoid $\alpha + \eta$ matrix.

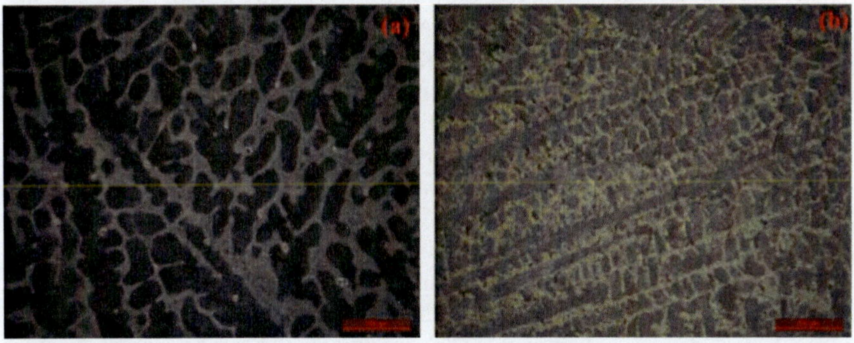

Fig. 11.1 Microstructure of ZA-27 alloy under a cooling rate of **a** 2.5 °C/min and **b** 10 °C/min

Increase in cooling rate from 2.5 to 10 °C/min resulted in refinement and change in morphology of the primary α phase. Further, the amount of eutectoid phase was found to be decreased with an increase in cooling rate. Unlike pure substances, the solidification of ZA-27 alloy is characterized by the presence of a mushy region (liquid + solid) between the fully solid and fully liquid regions in the solidifying domain. During solidification, the solute will be rejected from the mush zone into diffusion layer ahead of the dendritic front. The solute rejection and redistribution are affected by the composition of alloy (leads to concentration gradients) and cooling rate (leads to temperature gradient). Slow cooling during the solidification is expected to minimize the temperature gradient, facilitate homogenous distribution of solute, and hence promote the steady growth of primary solids. On the other hand, increased cooling during solidification is expected to maximize the temperature gradient and result in constitutional undercooling. Hence, it promotes more nucleation of primary solid and refinement of microstructure. ImageJ software is used to determine average arm size and phase fraction. The average arm size of 165.2 and 83.4 μm was obtained for ZA-27 alloy solidified under 2.5–10 °C/min. The amount of eutectoid fraction was nearly 35% and 23.5%, respectively, for ZA-27 alloy solidified under 2.5–10 °C/min. It was observed that an increase in cooling rate resulted in a change in the morphology of primary phase from irregular shape to nearly round shape.

Figure 11.2 shows microstructure of ZA-27 alloy modified with 1 wt% Nickel. The microstructure consists of heterogeneous structure of the primary α phase in eutectoid α + η matrix along with metallic compounds. Addition of nickel resulted in the rosette morphology of primary α phase. Nickel being a high melting element is expected to precipitate at the earlier stage of solidification and promotes nucleation of primary α phase. It results in more amount of primary α phase with rosette morphology and less amount of eutectoid phase. The amount of eutectoid phase was found to be 12.8% and 10.6%, respectively, for nickel modified ZA-27 alloy solidified under 2.5–10 °C/min. It was observed that the addition of nickel resulted in lower amount of eutectic but the change of cooling rate from 2.5 to 10 °C/min does not have any influence on the microstructure of ZA-27 alloy in presence of nickel. However, the morphology of metallic compound was found to be changed from irregular sharp plate/rectangular type to smooth round/cylinder type by an increase in cooling rate from 2.5 to 10 °C/min of nickel modified ZA-27 alloy. The XRD analysis was carried out to identify the phases as shown in Fig. 11.3. ZA-27 alloy solidified at 2.5 and 10 °C/min shows α-Aluminum and η-Zinc phases. The presence of the $MgZn_5$ phase was observed with ZA-27 alloy solidified at 2.5 °C/min. Samples cooled at a rate of 10 °C/min do not show peaks corresponding to $MgZn_5$. It indicates that the fast cooling during solidification prevents the crystallization of $MgZn_5$. Nickel-modified ZA-27 alloy shows metallic compound of Al_3Ni_5 phases. The comparison of relative intensities of XRD peaks corresponding to α and η phases shows that the addition of nickel resulted in a decrease in peak intensities of α phase and an increase in peak intensities of η phase. It confirms that the high melting point nickel (FCC) can easily combine with aluminum (FCC) during solidification and forms metallic compounds. It results in lowering the aluminum content in α phase and rejection of more Zn in η phase. The hardness of the samples was measured using Vicker's hardness machine at 5 kg load. Average hardness values of 107 HV_5 and 123 HV_5 were obtained for ZA-27 alloy solidified at the cooling rate of 2.5 and 10 °C/min, respectively. The corresponding values for nickel-modified alloy were 135 HV_5 and 172.5 HV_5. Increase in cooling rate resulted in the increase in hardness value due to the refinement of microstructure. Similarly, the addition of nickel resulted in an increase in hardness values due to the formation of metallic compounds.

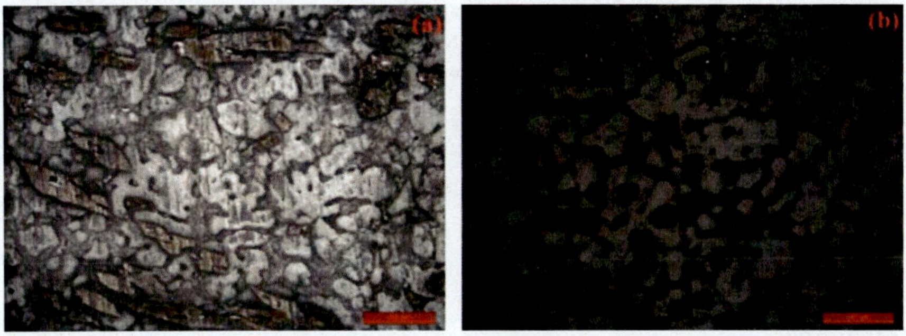

Fig. 11.2 Microstructure of Ni-modified ZA-27 alloy under a cooling rate of **a** 2.5 °C/min and **b** 10 °C/min

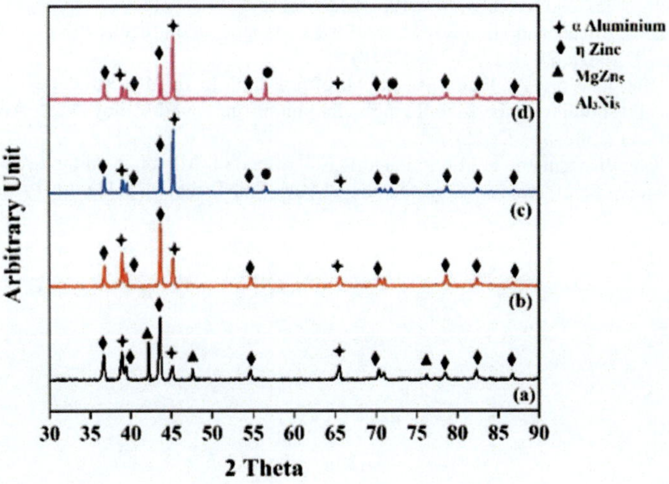

Fig. 11.3 XRD pattern of ZA-27 solidified at **a** 2.5 °C/min **b** 10 °C/min and nickel modified ZA-27 solidified at **c** 2.5 °C/min **d** 10 °C/min

11.4 Conclusions

The effect of the addition of nickel and cooling rate during solidification on microstructure evolution and hardness was investigated. Increase in cooling rate from 2.5 to 10 °C/min of ZA-27 alloy resulted in a decrease in average arm size of α phase from 165.2 to 83.4 μm and the morphology of the primary α phase changes from irregular shape to nearly round shape. Addition of 1 wt% nickel to ZA-27 alloy resulted in primary α phase with rosette morphology and less amount of eutectoid phase with metallic compound of Al_3Ni_5. Increase in cooling rate from 2.5 to 10 °C/min does not have any influence on the microstructure of ZA-27 alloy in the presence of nickel. However, the morphology of metallic compound was found to be changed from irregular sharp plate/rectangular type to smooth round/cylinder type. Hardness of ZA alloy was found to be increased with an increase in cooling rate and also addition of nickel.

References

1. Ramesh, G., Harihara Sudhan, K.R.C., Sreehari, P., Ramesh Kumar, S.: Influence of process parameters on solidification behavior of ZA-8 alloy. Mater. Today: Proc. **5**(1), 2726–2732 (2018)
2. https://www.eazall.com/technical-library. Last accessed 02 Oct 2022
3. Krupinski, M.: Effect of addition Ce on crystallisation kinetics and structure of Zn–Al–Cu alloys. Arch. Metall. Mater. **63**(3), 173–1178 (2018)
4. Yang, D., Xie, J., Wang, A., Wang, W.: Effects of rare earth La on microstructures, mechanical properties and sliding wear behavior of high-aluminium zinc foundry alloy ZA30. Appl. Mech. Mater. **117–119**, 360–363 (2011)
5. Ramesh, G., Vishwanath, H.M., Prabhu, K.N.: Effect of Mn on cooling behaviour and microstructure of chill cast Zn–Al (ZA8) alloy. Mater. Sci. Technol. **28**(11), 1301–1307 (2012)

Chapter 12
Influence of Molybdenum on Nitride Strengthening of Fe–Cr–Mo Alloys

S. Subrahmanyam, D. V. T. Pardhasaradhi, T. Savanth, and R. Sunil Kumar

Abstract It is important to understand the effect of molybdenum on the nitrided ferritic stainless steels as it enhances the interstitial diffusion of nitrogen in the ferrite matrix by controlling the precipitation of chromium nitrides and thus minimizes the amount of chromium depletion present in the stainless steels. In this study, Fe–16 wt%Cr–1 wt%Mo, and Fe–16 wt%Cr–2wt%Mo alloys, representative of certain ferritic stainless-steel compositions are prepared via powder metallurgy process route (mechanical alloying followed by compaction and sintering). These specimens are salt bath nitrided at 480 °C for a nitriding time period of 4 h. The phase transformations at the nitrided surfaces were investigated by XRD investigations. The hardness of these nitrided Fe–16%Cr–1%Mo and Fe–16%Cr–2%Mo alloys is measured using a Vickers hardness setup and compared with un-nitrided samples. It is found that an increase in molybdenum content enhanced the surface hardness of these nitrided alloys.

Keywords Fe–Cr–Mo alloys · Powder metallurgy route · Salt bath nitriding

S. Subrahmanyam (✉) · D. V. T. Pardhasaradhi · T. Savanth · R. S. Kumar
Department of Metallurgical & Materials Engineering, National Institute of Technology Andhra Pradesh, Tadepalligudem 534101, India
e-mail: subbuaugust1987@gmail.com

D. V. T. Pardhasaradhi
e-mail: saradhy555@gmail.com

T. Savanth
e-mail: savanthtadepalli@gmail.com

R. S. Kumar
e-mail: sunil.rajulapati@nitandhra.ac.in

A. Tewari et al. (eds.), *Proceedings of the 3rd International Conference on Advances in Materials Processing: Challenges and Opportunities*, Springer Proceedings in Physics 293, https://doi.org/10.1007/978-981-99-1971-0_12

12.1 Introduction

Powder metallurgical processing is a near-net shape technique that offers higher material utilization (>90%) compared to conventional casting methods, better microstructural homogeneity, and the ability to tailor novel compositions [1, 2]. The most commonly used alloying elements are the processing-friendly metals Cu, Ni, and Mo. Consequently, there is a strong desire in the PM industry to increase the use of less costly alloying elements. Cr is an attractive alternative since it, besides low cost, provides high hardenability and also recyclable components. The drawback is that Cr has a high affinity for oxygen, making oxidation and oxide reduction in PM processing of Cr-alloyed materials challenging [3]. Furthermore, the interaction between nitrogen and Cr-alloyed powder during processing is important to consider, since Cr also has high nitrogen affinity and is prone to form nitrides [4]. The diffusion behaviour of Cr and Mo during the sintering of Fe–Cr–Mo based alloys will help in estimating the improved density and hardness [5, 6].

Thermochemical treatments such as carburizing, boronizing and nitriding are widely implemented in various iron-based alloys and different types of steels, that will enhance mechanical, tribological and electrochemical properties by interstitial diffusion mechanisms which alter the surfaces harder and retain the bulk matrix soft and ductile [7–10]. The work reported in this paper discusses $Fe_{(84-x)}$–16wt%Cr–Mo_x alloys with (x = 1 and 2%), based on ferritic stainless-steel chemical composition and these alloys are fabricated through the conventional powder metallurgy route. The fabricated alloys are further salt bath nitrided to evaluate the surface properties.

12.2 Materials and Experiments

The elemental metal powders of Iron, Chromium, and Molybdenum with specifications as mentioned in Table 12.1 are used in this study. Mechanical alloying of these metal powders is carried out in a planetary ball mill (Retsch PM100, Germany), and tungsten carbide balls of 10 mm diameter are used. The mechanical alloying is performed for 25 g pre-mixed powders of corresponding $Fe_{(84-x)}$-16wt%Cr–Mo_x compositions separately with a rotational velocity of 300 rpm, with a ball-to-powder ratio of 10:1 for 5 h of milling time, with an interval of 15 min for each one hour of run time [11, 12]. These powders are further compacted using an oil hydraulic press with a load of 10 tons to prepare samples of 10 mm diameter and 4–5 mm height. These green compacts are sintered at 1200 °C for 90 min in an argon gas atmosphere, and then furnace-cooled [5].

These sintered Fe–Cr–Mo alloys prepared by powder metallurgy process route, are subjected to salt bath nitriding treatment at a temperature of 480 °C for 4 h using the cyanate salt bath composition as mentioned in Table 12.2.

Table 12.1 Elemental metal powders used in this study

Metal powder	Purity (%)	Particle size (microns)
Iron	99.9	45–55 μ
Chromium	99.0	45–55 μ
Molybdenum	99.99	45–55 μ

Table 12.2 Salt bath composition for liquid nitriding treatment [13]

Salt	Quantity (wt%)
Potassium thiocyanate (KOCN)	88
Potassium carbonate (K_2CO_3)	6
Sodium carbonate (Na_2CO_3)	3
Urea ($CO(NH_2)_2$)	3

An optical microscope was used for microstructure studies of the as-fabricated Fe–16%Cr–1%Mo and Fe–16%Cr–2%Mo alloys used in this study. Microhardness readings were taken at a load of 50 g for both (conditions) nitrided and un-nitrided, with a dwell time of 15 s. Patterns are recorded using an X-ray diffractometer (Rigaku Ultima III, Japan) from as-received and nitride surfaces of stainless-steel grades considered for this study, to identify the metal nitrides formed in the nitride regions of the specimens. Parameters considered for diffractometry include a scan rate of 2°/min with a step size of 0.02° and Cu anode with $k = 1.54056°Å$ for a scan range of 30–85°. The phases present in the as-received and metal nitrides formed in the liquid nitrided Fe–Cr–Mo alloys are identified using PANalytical Xpert High Score software and the developed phases were identified using the ICDD PDF-2 database [14].

12.3 Results and Discussion

12.3.1 Optical Microscopy

The Fe–Cr–Mo alloys prepared by ball milling and subsequent operations of the powder metallurgy process are prepared for optical microscope investigations by following standard metallography procedures. The polished specimens are etched with Ralph's agent for 15 s. Figure 12.1 shows the ferrite phase microstructures of Fe–16%Cr–1%Mo and Fe–16%Cr–2%Mo alloys as molybdenum and chromium are expected to dissolve in the α-ferrite matrix upon the mechanical alloying process. Subsequent sintering treatment of these alloys leads to further densification and strengthening by diffusion mechanism. However, the pores are still present in these alloys, which may be further controlled by adapting advanced sintering techniques such as hot pressing or spark plasma sintering.

Fig. 12.1 Microstructure of as-fabricated **a** Fe–16%Cr–1%Mo and **b** Fe–16%Cr–2%Mo alloys in powder metallurgy process route

12.3.2 XRD Analysis

The X-ray diffraction patterns of elemental molybdenum powder, mechanical alloyed and salt bath nitrided specimens of Fe–16%Cr–1%Mo and Fe–16%Cr–2%Mo alloys at 480 °C for nitriding time of 4 h are shown in Fig. 12.2. The metal nitride phases identified from the XRD analysis for nitrided Fe–16%Cr–1%Mo and Fe–16%Cr–2%Mo are Fe_2N, CrN. The amount of chromium nitride precipitation formed in Fe–16%Cr–2%Mo alloy upon nitriding is expected to decrease when compared to Fe–16%Cr–1%Mo alloy under similar nitriding conditions, as the molybdenum in stainless steel compositions will act as a substitute to chromium composition for maintaining similar hardenability of the steels. Also, from Fig. 12.2, it is understood that expanded ferrite phase formation upon the nitriding treatment of both Fe–16%Cr–1%Mo and Fe–16%Cr–2%Mo alloys is observed, with the α-phase peak shifting to lower diffraction angles.

12.3.3 Microhardness Measurements

The hardness values of the sintered compacts of Fe–16%Cr–1%Mo and Fe–16%Cr–2%Mo alloys and the same compacts upon subsequent salt bath nitriding treatment are shown in Fig. 12.3. It is evident that an increase in the molybdenum composition of the salt bath nitrided Fe–Cr–Mo alloys increased the surface hardness due to intermetallic compounds formed by molybdenum with chromium present in these alloys during sintering treatment.

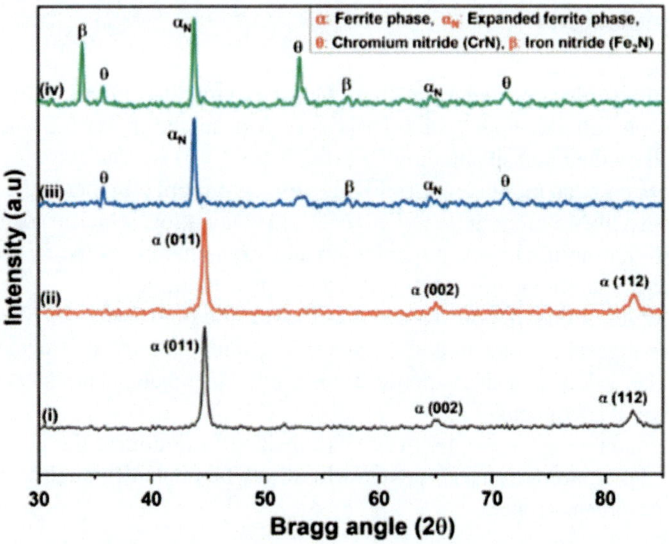

Fig. 12.2 XRD analysis of mechanically alloyed (i) Fe–16%Cr–1%Mo, (ii) Fe–16%Cr–2%Mo powders, and salt bath nitrided (iii) Fe–16%Cr–1%Mo and (iv) Fe–16%Cr–2%Mo alloys

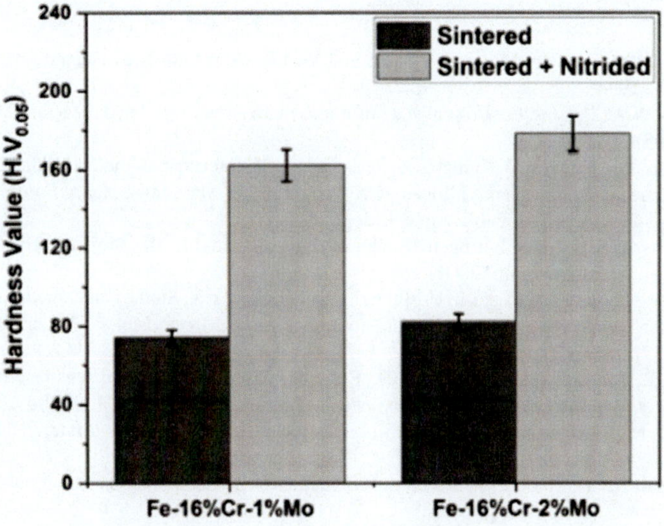

Fig. 12.3 Micro hardness values of Fe–16%Cr–1%Mo and Fe–16%Cr–2%Mo alloys before and after salt bath nitriding treatment

12.4 Conclusions

- It is found from XRD analysis that, Molybdenum dissolution in the ferrite matrix, is identified in both the Fe–16%Cr–1%Mo, and Fe–16%Cr–2%Mo alloys fabricated by the mechanical alloying process for 5 h of ball milling.
- It is understood that molybdenum alloying with chromium is higher in the sintered Fe–16%Cr–2%Mo alloy, than in Fe–16%Cr–1%Mo alloy, which leads to better control for chromium nitride precipitation and strengthening of the ferrite matrix at the surface, at the operated nitriding treatment conditions.
- Microhardness values of both Fe–16%Cr–1%Mo, and Fe–16%Cr–2%Mo sintered alloys are measured prior to and post-nitriding treatment and it is evident from the reported values that salt bath nitriding resulted in enhanced hardness to those of as sintered specimens.
- Advanced sintering techniques are recommended to minimize the porosity and achieve higher strength in the Fe–Cr–Mo alloys fabricated through the powder metallurgy process route.

References

1. German, R.M.: Powder Metallurgy Science. Metal Powder Industries Federation, Princeton, NJ, USA (1994)
2. Kaysser, W.A., Petzow, G.: Present state of liquid phase sintering. Powder Metall. **28**(3), 145–150 (1985)
3. Bergman, O., Lindqvist, B., Bengtsson, S.: Influence of sintering parameters on the mechanical performance of PM steels pre-alloyed with chromium. In: Materials Science Forum, vol. 534, pp. 545–548. Trans Tech Publications Ltd, (2007)
4. Bergman, O.: Effects of Nitrogen Uptake during Sintering on the Properties of PM Steels Prealloyed with Chromium (2001)
5. Chauhan, S., Verma, V., Prakash, U., Tewari, P.C., Khanduja D.: Studies on induction hardening of powder-metallurgy-processed Fe–Cr/Mo alloys. Int. J. Miner., Metall., Mater. **24**(8), 918–925 (2017)
6. Marcu, T., Molinari, A., Straffelini, G., Berg, S.: Microstructure and tensile properties of 3%Cr–0.5%Mo high carbon PM sintered steels, Powder Metall. **48**(2), 139 (2005)
7. Mozetič, M.: Surface modification to improve properties of materials. Materials **12**(3), 441 (2019)
8. Czerwinski, F.: Thermochemical treatment of metals. Heat Treatment–Conventional Nov. Appl. **26**(5), 73–112 (2012)
9. Mittemeijer, E.J., Marcel Somers, A.J.: Thermochemical Surface Engineering of Steels. Woodhead Publishing, Cambridge (2014)
10. Natarajan, S.: Thermochemical Surface Engineering of Steels, pp. 875–878 (2015)
11. Suryanarayana, C: Bibliography on Mechanical Alloying and Milling. Cambridge International Science Publishing (1995)
12. Park, J., Jeong, G., Kang, S., Lee, S.-J., Choi, H.: Fabrication of Fe–Cr–Mo powder metallurgy steel via a mechanical-alloying process. Met. Mater. Int. **21**(6), 1031–1037 (2015)

13. Kumar, N., Chaudhari, G.P., Meka, S.R.: Investigation of low-temperature liquid nitriding conditions for 316 stainless steel for improved mechanical and corrosion response. Trans. Indian Inst. Metals **73**(1), 235–42 (2020)
14. JCPDS. PDF-2 Database, Version 2.1. International Center for Diffraction Data (2002)

Chapter 13
Comparative Study of Heat Treatment of Ni–Cr–Mo Alloy Tubes Using Resistance and Induction Heating Furnaces

N. S. Dubey, B. Indranil, G. Das, B. Chandrasekhar, M. V. Ramana, G. Sugilal, Komal Kapoor, and Dinesh Srivastava

Abstract Ni–Cr–Mo based alloys have a clear advantage for high temperature application over the austenitic steels in terms of high creep resistance and high corrosion resistance. The fundamental challenges in manufacturing these thin-walled (D/t > 25) small diameter tubes of Ni–Cr–Mo based alloys are formation of α-Cr precipitate during heat treatment and high work hardening rate. Thermo-Mechanical Simulator studies, dilatometry studies and heat treatment trials in high temperature furnaces were carried out for understanding the reason for precipitation in the microstructure and establishing heat treatment parameters. The paper presents the work done towards establishing the heat treatment parameters for manufacturing these alloy tubes, the challenges faced and the methods undertaken to overcome them. The paper also covers a comparative study of heat treatment of Ni–Cr–Mo alloy tubes carried out in resistance and induction heating continuous furnaces.

Keywords Ni–Cr–Mo alloy · α-Cr precipitate · Resistance heating · Induction heating

13.1 Introduction

Ni–Cr–Mo based alloys have a clear advantage for high temperature application over austenitic steels [1]. The alloy also has high creep resistance and high corrosion resistance [2, 3]. High Cr content in the Ni–Cr–Mo based alloys, results in the

N. S. Dubey (✉) · G. Sugilal
Homi Bhabha National Institute, Mumbai, India
e-mail: ndubey@nfc.gov.in

N. S. Dubey · B. Indranil · G. Das · B. Chandrasekhar · M. V. Ramana · K. Kapoor · D. Srivastava
Nuclear Fuel Complex, Hyderabad, India

G. Sugilal
Bhabha Atomic Research Center, Mumbai, India

A. Tewari et al. (eds.), *Proceedings of the 3rd International Conference on Advances in Materials Processing: Challenges and Opportunities*, Springer Proceedings in Physics 293, https://doi.org/10.1007/978-981-99-1971-0_13

formation of brittle α-Cr phase and for dissolution of α-Cr phase, high temperature is required. However, at high temperatures grain coarsening is extremely fast. Thus, the work aimed to carry out microstructural studies of the Ni–Cr–Mo alloy to establish heat treatment process facilities and its parameter. Heat treatment of Ni–Cr–Mo alloy tubes are required to reduce hardness after cold working without solid precipitates. This paper summarizes the philosophy followed for the design of the furnace along with quenching facility, establishment of parameters for heat treatment of tubes, comparative studies, and conclusion.

13.2 Heat Treatment Furnaces

High temperature furnaces with a suitable atmosphere along with a quenching system is required to meet the heat treatment requirement. Heat treatment parameters such as temperature, time, quenching rate, etc. need optimisation based on tube size, chemical composition, and required mechanical and metallurgical properties.

A comparison of the phase diagram and dilatometry analysis reveals phase transformation of the Ni–Cr system above 1050 °C. Out of various options available for furnace atmospheres, the inert atmosphere was found to be most suitable. Accordingly, a vertical type experimental furnace was developed with an inert atmosphere and was found that 1100 °C and above is suitable for heat treatment due to the presence of Molybdenum in the alloy. Samples were heated in the experimental furnace and quenched at four different cooling rates i.e. 35, 62 and 68 °C/min and above 250 °C/min. The microstructure of the tube samples were analyzed, and an increase in grain size was observed with a decrease in cooling rate. Matrix showed Cr precipitate-free microstructure was obtained for all four cooling rates. However, a cooling rate of 250 °C/min and above were required to achieve an average grain size of ASTM 3 and better. Based on the above experimental results, heat treatment facility requirement has been divided into two types i.e. resistance and induction annealing furnace based on the tube size.

13.2.1 Resistance Roller and Annealing Furnace

Hot extruded blanks to final tubes are required to be heat treated in the temperature range of 1120–1140 °C followed by immediate quenching. The entire heat treatment requirement of the tubes is divided into two ranges (1) higher diameter i.e. 63 mm outer diameter (OD) to 21 mm OD and (2) smaller diameter i.e. 14 mm OD to 6 mm OD based on ratio of the surface area of material to be quenched to heat to be removed. To achieve rapid quenching in high wall thickness tubes, after soaking, water was selected as the quenching medium for a higher diameter tube and Argon for a smaller diameter tube. Accordingly, resistance heated horizontal tube roller furnace has been designed using SiC heaters and Fe–Cr–Al–Mo type rollers for higher diameter tubes and resistance heating continuous annealing furnace from the intermediate to the final size tube.

13.2.1.1 Resistance Roller Furnace

Resistance roller furnace consisted of inlet table, roller type furnace, outlet table and water quenching system as major systems. The loading table is meant for stacking of the tube and loading of the tube sequentially into the roller furnace. The roller furnace is designed for temperatures of 1200 °C. The heating capacity is calculated based on heat load, heating time and losses from the furnaces such as wall losses, flue gas losses, other losses, and safety factor. Multiple layers of thermal insulation of different grades are considered for reducing heat loss and skin temperature of the furnace. The thermal conductivity of the insulations was selected based on the manufacturer's catalog. The furnace internal convection coefficient was assumed to be very large such that T_1 is equal to the furnace temperature. In the case of the outside surface of the furnace, the convective heat transfer was considered to be a combination of forced and free convection. For this mixed convection condition, outer convection coefficient is calculated using Churchill [4] equations by calculating Nusselt, Rayleigh, Reynolds, and Prandtl numbers. The thermal resistance of each insulation was obtained using ASTM C680 [5]. The temperature profile is calculated at across the layers of insulations and found within acceptable limits of the insulation used (Fig. 13.1).

The furnace has five heating zones for better thermal uniformity. SiC heating element is used as it is suitable for application in temperature from 800 to 1600 °C in controlled atmospheres. Using surface loading curve, and a three phase star connected system, total heating elements were calculated and hot zone length was selected. To verify the element design, the maximum element temperature was calculated using Stefan Boltzmann equation and found below the maximum element temperature of

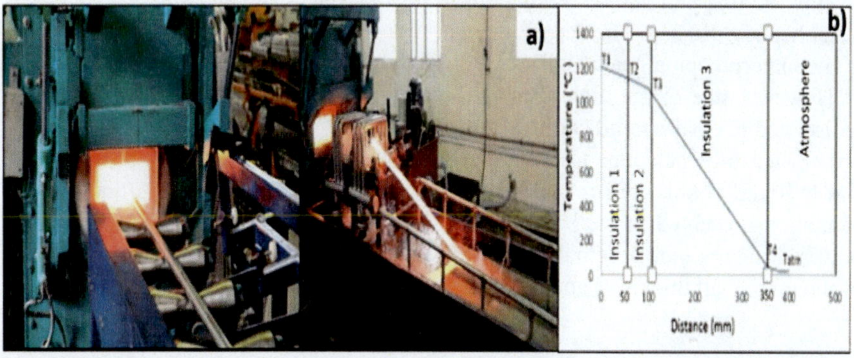

Fig. 13.1 **a** Inlet and outlet with quench tank of horizontal roller resistance heating furnace. **b** Temp. profile across insulation

SiC elements. Heated tube from the furnace is transferred with the help of outlet table transfers to the water quenching system.

In the vertical design experimental furnace, a temperature gradient was observed which could be minimized in the horizontal design. The tubes also had lesser bend and uniform oxide scaling. A temperature uniformity survey using a specially designed survey door and survey thermocouple was carried out. The temperature uniformity was mapped as per AMS 2750E and found to be within ±3 °C.

13.2.1.2 Resistance Annealing Furnace

The design of the hot zone of the Resistance Annealing furnace is similar to the roller furnace. Argon was selected as the quenching medium considering the low mass of tubes and the nearly oxidation-free surface of the tube after heat treatment.

A stacking table along with pinch rollers are used for loading the tube continuously into the furnace. The furnace is designed with three zones where the initial zone is for heating and the subsequent zones for uniform soaking. This is followed by a rapid cooling zone to ensure α-Cr precipitate free microstructure. However, the production rate in continuous annealing furnaces is much slower as productivity depends on the thermal conductivity of the material being heated.

13.2.2 Induction Annealing Furnace

To increase the productivity and improve the economics of the heat treatment during the intermediate to final stages of tubes, induction heating furnace is designed which enhanced up to six times higher output compared to the resistance heating continuous annealing furnace. A stacking table along with the tractor drive are used for loading the tube continuously into the furnace. Induction heating works on the principle of heating the material with an eddy current flowing inside the material and thus, is a faster heating process. The depth of penetration was considered at the wall thickness of the material being heated and accordingly, the heating frequency is selected [6].

However, due to the high speed of operation, the quenching system had to be redesigned to ensure rapid cooling with copper tubes and graphite sleeves. The base power had to be calculated for each size of the tube. This helped to minimize the end loss and obtain uniform recrystallized grains. The typical nature of induction heating is presented in the below graph (Fig. 13.2) depicting the temperature and power along the tube length during solution annealing due to change in appearance of the tube, wall thickness and diameter variation, etc.

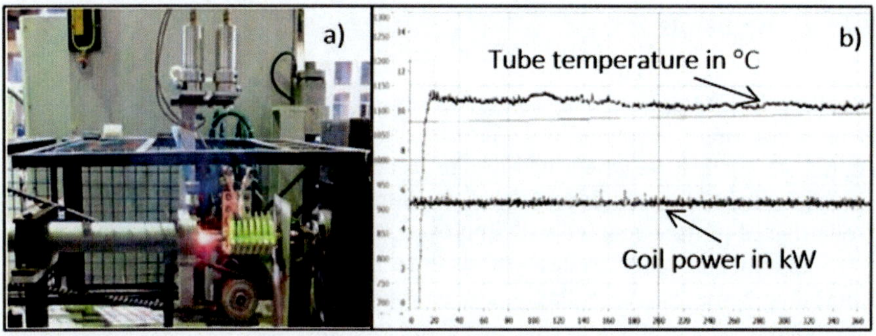

Fig. 13.2 **a** Induction annealing furnace **b** temperature and power profile along the length of the tube in induction annealing furnace

13.3 Results and Discussion

Performance evaluation study was carried out to compare the performance of induction and resistance heated furnaces for heat treatment of Ni–Cr–Mo alloy tubes. The microstructure of the heat-treated tube in the resistance furnace consisted of the equiaxed FCC austenite phase. The annealed microstructure appeared clean and did not show any sign of precipitation. Heat treatment in the horizontal roller furnace led to uniform oxide scaling on the tube without any appreciable bend as compared to the vertical design. This was also reflected in the metallographic properties (Table 13.1).

Minor fluctuations in the grain size of the induction heated material are observed which is unlikely in the case of resistance heated material. Grain size uniformity obtained was within the range of ASTM 6–10 while confirming other parameters. Thus, the present study reveals that induction heated furnace can be used in the initial stages of heat treatment of thicker tubes whereas the final stage of heat treatment can be carried out using resistance-heated annealing furnace to enhance productivity and product quality.

Table 13.1 Metallographic comparison of vertical, horizontal resistance furnace and induction furnace

Furnace type	Annealing parameter	Avg. grain (ASTM)	Mean hardness (HV 10)
Horizontal roller furnace	1120–1140 °C/WQ	4–5	180–190
Vertical annealing furnace		3–5	170–195
Induction annealing furnace	1120–1140 °C/AQ	6–10	190–210

13.4 Conclusion

- Solutionizing temperature and cooling rate of Ni–Cr–Mo alloy tubes is critical in ensuring precipitate-free microstructure. Vertical furnace for blanks and initial stages heat treatment had design restrictions leading to minor variation in microstructural properties. This was addressed by developing a horizontal roller-based furnace coupled with improved recovery and ease of operation.
- An induction furnace generates up to six times more output compared to resistance annealing furnaces. Accordingly, a double jacketed cooling chamber has been modified to achieve the requisite quenching rate. However, an induction furnace is more suitable for intermediate sizes of the tubes as compared to final pass tubes.

References

1. Kirchner, R.W., Hodge, F.G.: New Ni–Cr–Mo alloy demonstrates high-temperature structural stability with resultant increases in corrosion resistance and mechanical properties. Mater. Corros. **24**, 1042–1049 (1973)
2. Mishra, A., Richesin, D., Rebak, R.B.: Localized Corrosion study of Ni–Cr–Mo Alloys for Oil and Gas Application. NACE International, paper No. 5802 (2015)
3. Friend, W.Z.: Corrosion of Nickel and Nickel-Based Alloys, pp. 292–431. John Wiley and Sons, New York, NY (1980)
4. Churchill, S.W.: Combined Free and Forced Convection Around Immersed Bodies Heat Exchanger Design Handbook. Hemisphere Publishing Corp., New York, NY (1983)
5. ASTM-C680.: Standard Practice for Estimate of the Heat Gain or Loss and the Surface Temperatures of Insulated Flat, Cylindrical, and Spherical Systems by Use of Computer Program. Annual Book of ASTM Standards, vol. 04.06, West Conshohocken, PA, United States (1989)
6. Rudnev, V., Loveless, D., Cook, R., Black, M.: Handbook of Induction Heating, 2nd edn. CRC Press, London (2017)

Chapter 14
Electron Beam Welding and Gas Tungsten Arc Welding Studies on Commercially Pure Titanium Sheets

Vaibhav Gaur, R. K. Gupta, and V. Anil Kumar

Abstract Commercially pure titanium (CP-Ti) sheets have been welded through gas tungsten arc welding (GTAW) using Ti6Al4V-ELI filler wire and through electron beam welding (EBW). Microhardness is found to be higher in the weld fusion zone as compared to the parent metal. The strength of the weldments of GTAW and EBW are found to be nearly same. Although GTAW weldments are expected to exhibit lower strength compared to that of EBW, the use of Ti6Al4V-ELI filler wire contributed to a higher strength in the GTAW weldment. Failure of the tensile test specimens was found to occur in the parent metal indicating the weld efficiency ~100% in both the cases. However, a marginal reduction in ductility of welded samples is observed. Widmanstätten microstructure is observed in the weld fusion zone of both the weldments. The microstructure is found to be relatively finer in EBW which is attributed to the lower heat input and higher cooling rate as compared to that in GTAW. Coarse and fine dimples are observed in the fractographs of the tensile tested samples indicating good ductility in both the weldments.

Keywords CP-Ti · Ti6Al4V-ELI filler wire · EBW · GTAW · Weld efficiency

14.1 Introduction

Commercially pure titanium (CP-Ti) is used extensively in marine, chemical, defense and aerospace sectors due to its superior specific strength and high corrosion resistance. It is generally used for fabrication of components for cryogenic applications and gas bottles in aerospace systems [1]. It exhibits good amount of ductility and

V. Gaur (✉)
Academy of Scientific and Innovative Research (AcSIR), Ghaziabad 201002, India
e-mail: vaibhav.nml19a@acsir.res.in

CSIR-National Metallurgical Laboratory, NML P.O., Jamshedpur 831007, India

R. K. Gupta · V. A. Kumar
ISRO-Vikram Sarabhai Space Centre, Trivandrum 695022, India

A. Tewari et al. (eds.), *Proceedings of the 3rd International Conference on Advances in Materials Processing: Challenges and Opportunities*, Springer Proceedings in Physics 293, https://doi.org/10.1007/978-981-99-1971-0_14

hence sheets are used in the fabrication of components. Oxygen is an interstitial alloying element, which is also an α-stabilizer and provides strength to the material [2]. At high temperature, titanium reacts with the atmospheric gases which affect the properties and structure of the material. This also makes welding activity as a challenging task. The welding techniques which are most commonly used are electron beam welding (EBW) and gas tungsten arc welding (GTAW). Apart from these welding techniques, friction stir welding (FSW), laser beam welding (LBW) and resistance welding are also used [3].

EBW is considered as the best among these due to exhibition of high weld efficiency, repeatability and clean operation [4]. It is a single pass welding which provides the joint with narrow width and with deeper penetration along with less heat affected zone [5]. One limitation in EBW is the size of the vacuum chamber, which limits the size of the components that can be welded. In such cases, GTAW can be employed with use of shielded gases both from the top and bottom sides. EBW and GTAW (using filler wire of CP-Ti) for thin sheet welding has been reported [6]. The microstructure of the parent metal present initially in the material gets modified at the weld joints and in turn affects the mechanical properties. GTAW has been carried out with Ti6Al4V-ELI filler wire to mitigate these effects. However, limited literature is available in this area and hence the present work would provide insights into this area.

14.2 Experimental Procedure

5 mm thick sheet of CP-Ti was cut into 3 samples of size 125 mm × 50 mm × 5 mm (with rolling direction parallel to the length) and were further cut into two halves for preparing the weld coupons for EBW and GTAW. For EBW, a square-grooved butt joint was produced whereas for GTAW Y-shaped butt joint was chosen with angle between both halves as 60° up to a depth of 3 mm. EB welding (voltage 60 kV, current 110 mA and welding speed 30 mm/s) was done in a single pass. For GTAW, welding was carried out with Ti6Al4V-ELI filler wire of 1.6 mm diameter in 3 passes, 2 pass from the face side and one pass from the root side (voltage 11.5 V, current 90 A and argon gas flow rate 15 L/min). Weld coupons were subjected to non-destructive testing for ensuring defect free welds. Optical microscopy of the weldment cross-section was done using Olympus make GX-11 optical microscope and microhardness survey of welds was done at an interval of 0.5 mm using Qness-Q10 A+ microhardness tester. The tensile testing was conducted at a constant cross-head movement rate of 2 mm/min. Fractography of tensile tested samples was done using Zeiss make EV050 scanning electron microscope (SEM).

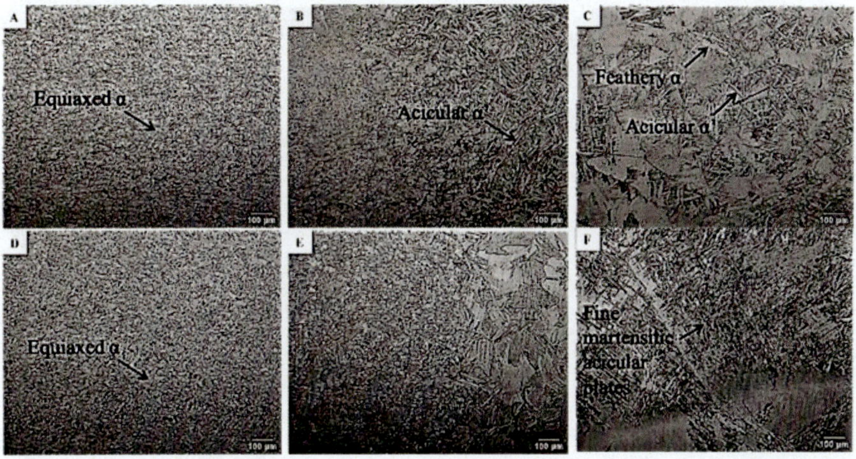

Fig. 14.1 Microstructure of EBW and GTAW joints of CP-Ti, **a**, **d** parent metal, **b**, **e** HAZ-weld interface and **c**, **f** weld fusion zone respectively

14.3 Results and Discussion

14.3.1 Microstructure of Welded Joints

Optical microscopy of both the weldments was carried out (Fig. 14.1). It can be seen that the microstructure of the parent metal consists of fine equiaxed α grains (Fig. 14.1a, d) representative of annealed condition. The interface region of both the weldments is shown in Fig. 14.1b, e. In case of EBW, weldment microstructure (Fig. 14.1c) consists of acicular α′ along with feathery α and fine prior β grains. As the solidification of EBW melt pool does not take place by epitaxial mode due to high cooling rate, the parent metal has less effect on the microstructure of the weld and a fine acicular structure resultant of critical cooling for fine martensitic formation is observed. During the solidification of the GTAW weldment, epitaxial growth at the fusion boundary occurs and equiaxed grains at the center of the weld develop.

14.3.2 Microhardness

The microhardness survey of the weldments of EBW and GTAW at an interval of 0.5 mm is shown in Fig. 14.2. Microhardness in case of GTAW is higher than that of EBW and is attributed to the use of Ti6Al4V-ELI filler wire.

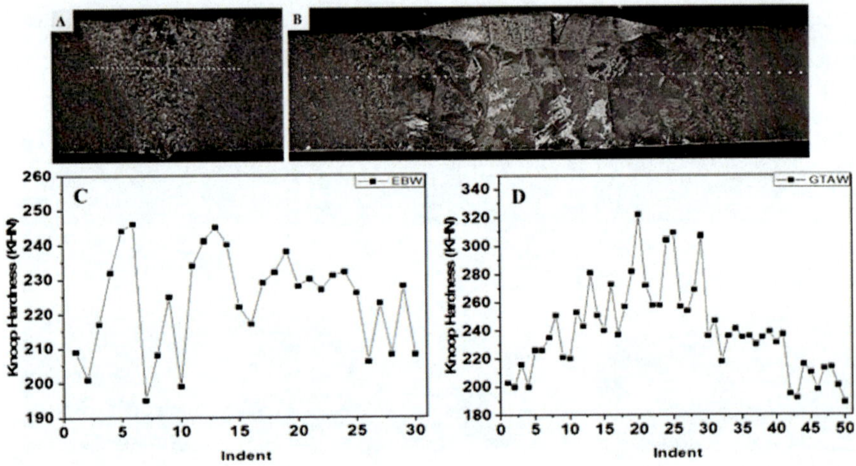

Fig. 14.2 Microstructure and microhardness plots of EBW (**a**, **c**) and GTAW (**b**, **d**) CP-Ti coupons

14.3.3 *Tensile Properties*

The tensile properties of EBW, GTAW and parent metal are shown in Table 14.1. The failure occurs in all the cases in the parent metal which indicates that the weld fusion zone is stronger than the parent metal. The weld efficiency w.r.t. both UTS and YS in both the weldments is ~100%. In both the cases, the dual necking is observed in the parent metal on both the sides of the weld which results in the failure from the parent metal. This clearly shows that the strength in case of welded specimens is higher than that of the parent metal. Generally, the strength in the case of GTAW is less than the parent metal, but the higher strength in the present case can be attributed to the presence of higher alloy content of Al and V by the virtue of use of alloy (Ti6Al4V-ELI) filler wire. % Elongation in case of the welded material is reduced with the increasing strength of the weld. The % elongation of the parent metal is ~32% which gets reduced to around 25 and 18% in case of EBW and GTAW, respectively, due to weldment microstructure in both the cases and additionally the effect of alloyed filler wire in GTAW weldment.

Table 14.1 Tensile test results of parent metal, EBW and GTAW samples of CP-Ti

	UTS (MPa)	YS (MPa)	Elongation (%)	Necking observation	Failure location
Parent metal	572 ± 5	477 ± 6	32 ± 0.9	Single necking	Parent metal
EBW	608 ± 3	494 ± 3	25.5 ± 0.4	Dual necking	Parent metal
GTAW	615 ± 5	506 ± 4	18.1 ± 0.6	Dual necking	Parent metal

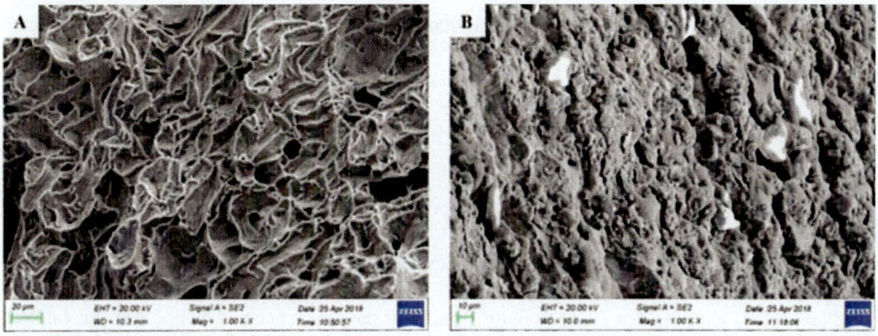

Fig. 14.3 Fractographs of tensile tested samples of CP-Ti **a** EBW and **b** GTAW

14.3.4 Fractography

Fractography analysis of the tensile tested specimens of weldments has been done to identify the mode of failure as shown in Fig. 14.3. From Fig. 14.3a it is seen that the fine dimples along with coarser dimples are present on the fractured surface of the tensile tested samples. This shows that there is a sustainable amount of deformation occurring in the material for EBW. In case of GTAW (Fig. 14.3b) it can be seen that the dimples are finer indicating limited deformation within the material before the fracture.

14.4 Conclusions

CP titanium sheets have been welded through EBW and GTAW welding techniques and mechanical properties at ambient temperature have been evaluated. The conclusions from the present study are as follows.

1. Location of the fracture in all the specimens is in the parent metal along with the dual necking on both the sides of the weldment indicating that weldment is stronger than the parent metal. Weld efficiency of both the welds is ~100% confirming the effect of use of Ti6Al4V-ELI filler wire in GTAW.
2. Higher strength in the case of GTAW compared to EBW and the parent metal is mainly attributed to the alloy content of Al and V in the weldment due to the use of Ti6Al4V-ELI filler wire.
3. The microhardness of the weldments decreased from the location of fusion zone to the HAZ to the parent metal. Also, the microhardness of GTAW is higher than that of EBW due to the use of alloy filler wire.
4. %Elongation of the GTAW is lower than that of EBW which can be attributed to the weldment microstructure and Al, V content in the GTAW weldment of CP-Ti.

Acknowledgements The authors would like to thank Dy Director MME and Director, VSSC, for permission to publish this work.

References

1. Boyer, R.R.: An overview on the use of titanium in the aerospace industry. Mater. Sci. Eng., A **213**(1–2), 103–114 (1996)
2. Williams, J.C.: Kinetics and phase transformation. Titanium science and technology. In: Jaffee, R.I., Burte H.M. (eds.) Second International Conference the Metallurgical Society of AIME 1973, vol. 3, pp. 1433–1494. Plenum Press, London (1973)
3. Webster, R.T.: Welding of titanium alloys. Soldering. In: Olson, D.L., Siewert, T.A., Liu, S., Edwards G.R. (eds.) ASM Handbook, vol. 6, pp. 783–786. ASM International, Metals Park (1993)
4. Wei, L., Shi, Y., Lei, Y., Li, X.: Effect of electron beam welding on the microstructures and mechanical properties of thick TC4-DT alloy. Mater. Des. **34**, 509–515 (2012)
5. Bing, W., Jinwei, L., Zhenyun, T.: Study on the electron beam welding process of ZTC4 titanium alloy. Rare Met., Mater. Eng. **43**(4), 786–790 (2014)
6. Gupta, R.K., Kumar, V.A., Xavier, X.R.: Mechanical behavior of commercially pure titanium weldments at lower temperatures. J. Mater. Eng. Perform. **27**(5), 2192–2204 (2018)

Part II
Advanced Steel Technology

Chapter 15
Hot Deformation Behavior of Medium Carbon Low Alloy Steel Using Arrhenius and ANN Modeling Methods

Hafeez Shekh, Sumit Kumar, and Sumeer K. Nath

Abstract Isothermal hot compression tests on medium carbon low alloy steel were carried out over the temperature range from 800 to 1050 °C and the strain rates from 0.001 to 0.1 s^{-1}. The hot deformation behavior of the tested steel is expressed in terms of activation energy and material constants using a universal constitutive equation. The present study explains the kinetics of dynamic recrystallization (DRX) using Avrami-DRX-model of the subject steel. Moreover, the comparative study has been performed between the ANN model and the Arrhenius constitutive model for the prediction of flow stress. The correlation coefficient between experimental flow stress and the predicted stress by the ANN model and Arrhenius model were obtained as 0.996 and 0.978, respectively. And the Avrami-DRX model for subject steel concluded as $X_{DRX} = 1 - \exp\left(-0.528\left(\frac{\varepsilon-\varepsilon_c}{\varepsilon_s-\varepsilon_p}\right)^m\right)$, $m = 24.64 - 0.80 * \ln(Z)$.

Keywords Dynamic recrystallization · Artificial neural network · Hot deformation

15.1 Experimental Condition

15.1.1 Steel Used

The present study is conducted on medium carbon low alloy steel of chemical composition (wt%) shown in Table 15.1.

H. Shekh (✉) · S. Kumar · S. K. Nath
Department of Metallurgical and Materials Engineering, Indian Institute of Technology Roorkee, Roorkee, Uttarakhand 247667, India
e-mail: hafeez.shekh@mt.iitr.ac.in

A. Tewari et al. (eds.), *Proceedings of the 3rd International Conference on Advances in Materials Processing: Challenges and Opportunities*, Springer Proceedings in Physics 293, https://doi.org/10.1007/978-981-99-1971-0_15

Table 15.1 Chemical composition of medium carbon low alloy steel (wt%)

C	Mn	Si	S	P	Cr	Fe
0.22	0.59	0.04	0.03	0.02	0.01	Bal.

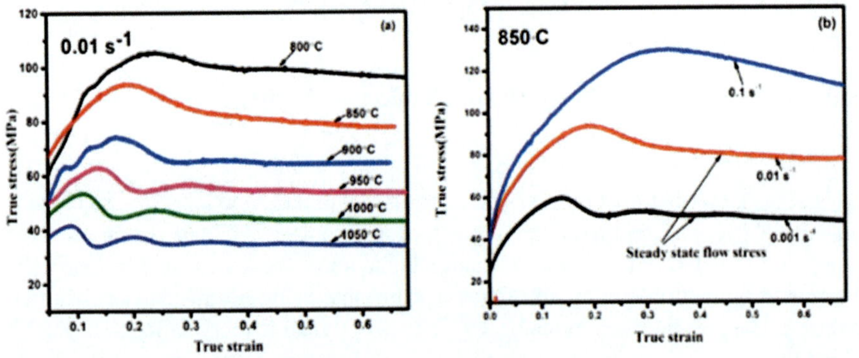

Fig. 15.1 Stress–strain curves for **a** constant 0.01 s^{-1} strain rate with varying deformation temperature, **b** constant deformation temperature of 850 °C with varying strain rate

15.2 Results and Discussion

15.2.1 Analysis of Flow Stress Curves

The experimental stress–strain curves for the subject steel are shown in Fig. 15.1. It can be observed from Fig. 15.1a, b that with increase in isothermal temperature and decrease in the strain rate the rate of dynamic softening increases and the corresponding peak stress, peak strain, and steady-state stress decrease. It is obvious that with an increase in temperature and decrease in the strain rate peak of the stress shifted towards the lower side due to the predominant process of DRV and DRX which can be confirmed with Fig. 15.1.

15.2.2 Arrhenius Model

It can be seen from Fig. 15.1 that flow stress varies with the variation of strain rate and deformation temperature. Zener–Hollomon parameter (Z) combines the effect of temperature and strain rate [1–5] and same has been used in the present study for subject steel.

$$Z = \dot{\varepsilon} \exp\left(\frac{Q}{RT}\right) \tag{15.1}$$

Based on the stress level, below are the relation between the Z-parameter and flow stress

$$Z = \dot{\varepsilon}\exp\left(\frac{Q}{RT}\right) = A_p\sigma^{n'} \quad (\text{Power law, for } \alpha\sigma < 0.8) \tag{15.2}$$

$$Z = \dot{\varepsilon}\exp\left(\frac{Q}{RT}\right) = A_E\exp(\beta\sigma) \quad (\text{Exponential law, for } \alpha\sigma > 1.2) \tag{15.3}$$

$$Z = \dot{\varepsilon}\exp\left(\frac{Q}{RT}\right) = A_h[\sinh(\alpha\sigma)]^n \quad (\text{Hyperbolic law, for all values of } \sigma) \tag{15.4}$$

where, Q and R, are the apparent activation energy and the universal gas constant respectively. A_E, A_p, A_h, n', β and n are material constants, and $\alpha \approx \beta/n'$ is an adjustable constant that brings $\alpha{*}\sigma$ into the correct range. To find the value of α, a method proposed by Uvira and Jonas [5] was used in this study. According to Sellars et al. [6], among the above three equations, hyperbolic-sine (Eq. (15.4)) is more general and valid over the wide range of stress level. To find the material constants and the activation energy the steps mentioned in reference [1–5] have been used and the calculated values of the material constants n, β,n', A_h, α and Q (kJ/mol) at true strain of 0.6 are 3.27, 0.091, 4.91, 12.132E+8, 0.018 and 263.71 respectively. After incorporating these material constants into Eq. (15.4) the hot deformation behavior of the medium carbon low alloy steel at true strain of 0.6 can be modeled as follows;

$$Z = \dot{\varepsilon}\exp\left(\frac{263710}{RT}\right) = 12.132 * 10^8 * [\sinh(0.018 * \sigma)]^{3.27} \tag{15.5}$$

$$\sigma = \frac{1}{0.018}\ln\left[\left(\frac{Z}{12.132 * 10^8}\right)^{\frac{1}{3.27}} + \left\{\left(\frac{Z}{12.132 * 10^8}\right)^{\frac{2}{3.27}} + 1\right\}^{1/2}\right] \tag{15.6}$$

15.2.3 Kinetics of DRX and SEM Micro Structure Analysis

To determine the extent of dynamic recrystallization, Avrami-DRX model [7] in general form can be represent by Eq. (15.7)

$$X_{DRX} = 1 - \exp\left(-k(f(\varepsilon))^m\right) \tag{15.7}$$

where; X_{DRX} represent a fraction of dynamic recrystallization, m and k are the material dependent terms and $f(\varepsilon)$ is the strain dependent function. Mirzaee et al. [8] have used the term $\left(\frac{\varepsilon-\varepsilon_p}{\varepsilon_s-\varepsilon_p}\right)$ as strain dependent function for low carbon low alloy steel. This represents DRX starts when the true strain reaches the peak strain and continues till the steady-state. Zhang et al. [9] have used the term $\left(\frac{\varepsilon-\varepsilon_c}{\varepsilon_p}\right)$ as the strain dependent function for medium carbon Cr–Ni–Mo alloyed steel which considers the start of DRX at an inflection point on θ–σ plot. In the previous section we have shown that DRX starts at an inflection point and completes its DRX cycle at steady-state strain. Hence, the use of $\left(\frac{\varepsilon-\varepsilon_c}{\varepsilon_s-\varepsilon_p}\right)$ as the strain dependent function for medium carbon low alloy steel would be a good agreement. After considering the mentioned strain function the Eq. (15.7) Avrami-DRX equation can be written as Eq. (15.8).

$$X_{DRX} = 1 - \exp\left(-k\left(\frac{\varepsilon - \varepsilon_c}{\varepsilon_s - \varepsilon_p}\right)^m\right) \tag{15.8}$$

According to Mirzaee et al. [8] the X_{DRX} can also be inferred as the ratio of flow stress drops from the peak stress $(\sigma_P - \sigma)$ to the whole softening $(\sigma_s - \sigma_P)$. After combining this interpretation into the above equation and plotting the linear fit between $\ln\{\ln(1/(1 - X_{DRX})\}$ and $\ln\{(\varepsilon - \varepsilon_c)/(\varepsilon_s - \varepsilon_p)\}$ the average value of k and the relation between m and Z-parameter are calculated as follows, $k_{\text{avg.}} = 0.528$ and $m = 24.64 - 0.80 * \ln(Z)$.

After substituting the values of m and k in Eq. (15.8) the recrystallization graphs are obtained as shown in Fig. 15.2 (**2.1**). It is observed from the stress–strain curves in Fig. 15.1, all the flow curves (except 850 °C and 0.1 s^{-1}) exhibit a peak followed by flow softening and then achieve steady-state stress which delineates the occurrence of DRX. At 850 °C and 0.1 s^{-1}, the steady-state stress was less apparently achieved which may show partial DRX (PDRX) [10]. Figure 15.2 (**2.2**) depicts the recrystallized grains at temperature (900, 950 °C) and strain rate 0.01 s^{-1} which confirms the X_{DRX} prediction by Eq. (15.8).

15.2.4 Prediction of Flow Stress by Artificial Neural Network (ANN) Model

For prediction of flow stress, Levenberg–Marquardt (L–M) training algorithm is used which uses the Jacobian function for calculation, and this algorithm is based on a feed-forward back propagation method.

The artificial neural network has been formed by random selection of 820 data points from the stress–strain data set. These 820 data points were further divided into three subsets namely, training (70%), testing (15%), and validation (15%) data sets. Then, these data sets normalized between 0 and 1 by Eqs. (15.9) and (15.10) [10].

$$\acute{y} = 0.1 + 0.8\left(\frac{y-y_{min}}{y_{max}-y_{min}}\right) \tag{15.9}$$

$$\acute{\dot{\varepsilon}} = 0.1 + 0.8\left(\frac{ln\dot{\varepsilon}-ln\dot{\varepsilon}_{min}}{ln\dot{\varepsilon}_{max}-ln\dot{\varepsilon}_{min}}\right) \tag{15.10}$$

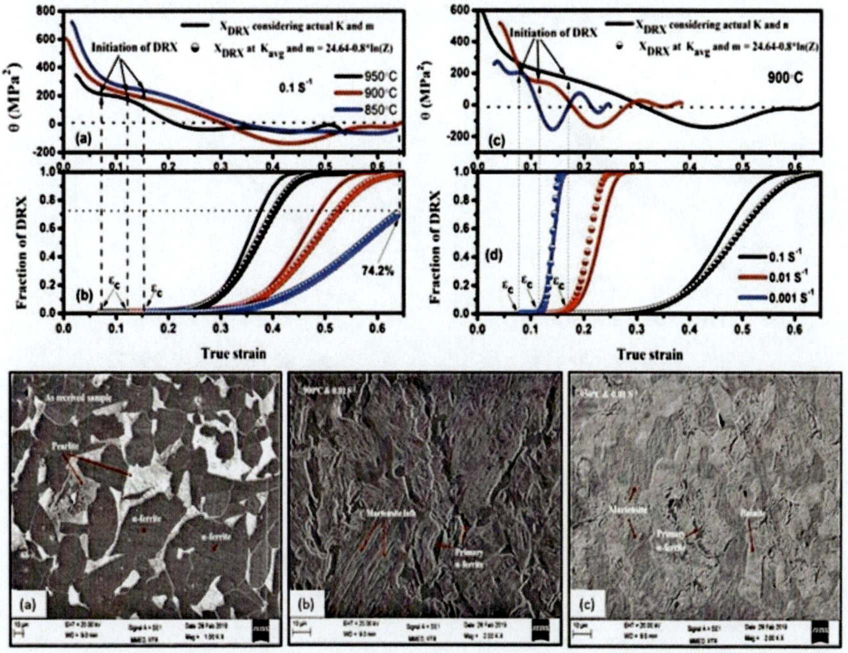

Fig. 15.2 (**2.1**) Plot of θ versus strain at different deformation conditions (**a**, **c**), fraction of DRX with strain (**b**, **d**). (**2.2**) SEM micrographs of the as received and hot deformed sample at different hot deformation conditions (**a**) as received sample, (**b**) 900 °C and 0.01 s^{-1}, (**c**) 950 °C and 0.01 s^{-1}

The performance of an artificial neural network for all the stages (training, testing, and validation) is shown in Fig. 15.3 (**3.1**). It can be clearly understood from Fig. 15.3 (**3.1**) (a–c) that all the predicted data points almost have exact match with experimental data points and the correlation coefficient for all the stages are close to one.

The strain ranging from 0.1 to 0.65 with an increment of 0.05 has been chosen to compare predicted stress by both Arrhenius and ANN models with the experimental stress. It is obvious from Fig. 15.3 (**3.2**) that the predicted stress by ANN model has better propensity to trace the experimental flow stress than the Arrhenius model.

15.3 Conclusion

1. Calculated activation energy at 0.6 true strain was 263.71 kJ/mol and Arrhenius model for the flow stress calculation is $Z = \dot{\varepsilon}\exp\left(\frac{263710}{RT}\right) = 12.132 * 10^8 * [\sinh(0.018 * \sigma)]^{3.27}$.
2. Avrami-DRX model for medium carbon low alloy steel can be expressed as; $X_{DRX} = 1 - \exp\left(-0.528\left(\frac{\varepsilon-\varepsilon_c}{\varepsilon_s-\varepsilon_p}\right)^m\right)$, $m = 24.64 - 0.80 * \ln(Z)$.

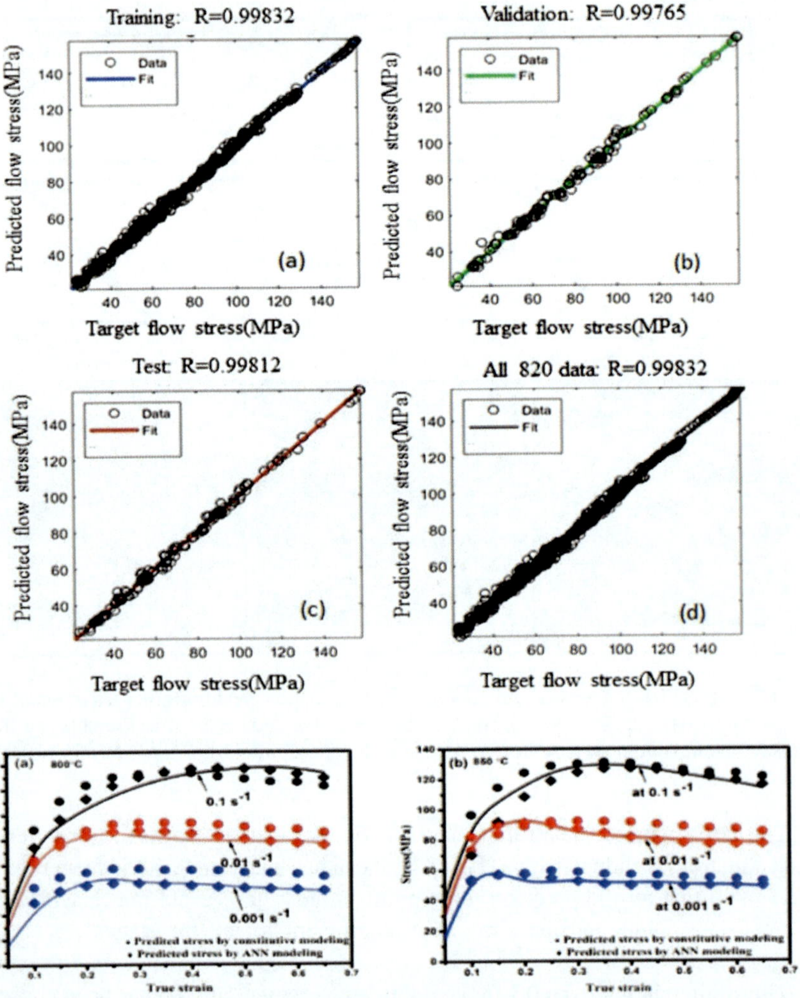

Fig. 15.3 (**3.1**) Linear fitting between predicted and targeted flow stress during; (**a**) training stage, (**b**) validation stage, (**c**) testing stage, (**d**) combine stages. (**3.2**) Comparative stress–strain curves; (**a**) at 800 °C, (**b**) at 850 °C

3. The ANN model has more propensity to trace the experimental flow stress than the Arrhenius model, thus the ANN model can better predict the flow stress.

References

1. Rajput, S.K., Chaudhari, G.P., Nath, S.K.: Physical simulation of hot deformation of low-carbon

Ti–Nb microalloyed steel and microstructural studies. J. Mater. Eng. Perform. **23**, 2930–2942 (2014). https://doi.org/10.1007/s11665-014-1059-8

2. Rajput, S.K., Chaudhari, G.P., Nath, S.K.: Characterization of hot deformation behavior of a low carbon steel using processing maps, constitutive equations and Zener–Hollomon parameter. J. Mater. Process. Technol. **237**, 113–125 (2016). https://doi.org/10.1016/j.jmatprotec.2016.06.008
3. Kumar, N., Kumar, S., Rajput, S.K., Nath, S.K.: Modelling of flow stress and prediction of workability by processing map for hot compression of 43CrNi steel. ISIJ Int. **57**, 497–505 (2017). https://doi.org/10.2355/isijinternational.ISIJINT-2016-306
4. Kumar, S., Chaudhari, G.P., Nath, S.K.: Critical conditions for dynamic recrystallization of hot deformed 1 wt% chromium–1 wt% molybdenum rotor steel. Mater. Perform. Charact. **8**, 20190022 (2019). https://doi.org/10.1520/mpc20190022
5. Uvira, J.L., Jonas, J.J.: Hot compression of Armco iron and silicon steel. Trans. Metall. Soc. AIME **242**, 1619–1926 (1968)
6. Sellars, C.M., McTegart, W.J.: On the mechanism of hot deformation. Acta Metall. **14**, 1136–1138 (1966)
7. Sheikh, H., Serajzadeh, S.: Estimation of flow stress behavior of AA5083 using artificial neural networks with regard to dynamic strain ageing effect. J. Mater. Process. Technol. (2008)
8. Mirzaee, M., Keshmiri, H., Ebrahimi, G.R., Momeni, A.: Dynamic recrystallization and precipitation in low carbon low alloy steel 26NiCrMoV 14-5. Mater. Sci. Eng., A **551**, 25–31 (2012)
9. Zhang, C., Zhang, L., Shen, W., Liu, C., Xia, Y., Li, R.: Study on constitutive modeling and processing maps for hot deformation of medium carbon Cr–Ni–Mo alloyed steel. Mater. Des. **90**, 804–814 (2016)
10. Singh, K., Rajput, S.K., Mehta, Y.: Modeling of the hot deformation behavior of a high phosphorus steel using artificial neural networks. Mater. Discov. **6**, 1–8 (2016). https://doi.org/10.1016/j.md.2017.03.001

Chapter 16
Comprehending Structure–Property Relationship in Hot-Rolled Low Alloy Steels

G. K. Bansal, A. K. Chandan, Chiradeep Ghosh, V. Rajinikanth, V. C. Srivastava, and S. Ghosh Chowdhury

Abstract In the present work, two newly designed low alloy steels were processed through the hot rolling and water quenching process at room temperature. The XRD and microstructural investigations suggest the presence of a multiphase microstructure containing various combination of martensite, ferrite, retained austenite and carbides. Between both the alloys, the presence of ferrite could be eliminated in the alloy with higher carbon and Mn content (Alloy-2). The presence of carbides and a small amount of retained austenite, even during fast water quenching to room temperature, can be attributed to the faster carbon diffusion due to its smaller size. The presence of higher carbon and Mn content in the Alloy-2 has resulted in higher hardness and strength with a slight decrease in the ductility, when compared with Alloy-1. Both the alloys showed superior mechanical properties, i.e. ultra-high strength (~900–1100 MPa YS and ~1400–1600 MPa UTS) combined with sufficient ductility (~11–14% total elongation). As a result, the energy absorption capability, which is the product of UTS and total elongation, was also measured to be on the higher side (~18–20 GPa%).

Keywords Low alloy steel · Martensite · Carbides · Strength · Ductility

16.1 Introduction

Steel is a predominant material that can be used in various manmade structures and machine components. This is because of its ease of recycling, capability for mass manufacturing, cost-effectiveness, etc. [1]. All the steel grades necessarily contain iron with up to 2.1 wt% carbon, however, some other alloying elements, namely Mn,

G. K. Bansal (✉) · A. K. Chandan · V. Rajinikanth · V. C. Srivastava · S. G. Chowdhury
Materials Engineering Division, CSIR-National Metallurgical Laboratory, Jamshedpur 831007, India
e-mail: bansal.gk@nmlindia.org

C. Ghosh
Research and Development Division, Tata Steel Limited, Jamshedpur 831001, India

A. Tewari et al. (eds.), *Proceedings of the 3rd International Conference on Advances in Materials Processing: Challenges and Opportunities*, Springer Proceedings in Physics 293, https://doi.org/10.1007/978-981-99-1971-0_16

Cr, Ni, Al, Si, etc. are also added to achieve the properties/performance required for a particular application. The capacity to withstand the specified load, performance against corrosion and wear, resistance to fracture, ease of bulk manufacturing, cost, etc. are main critical factors considered to decide the suitability of a steel grade for a particular application. Therefore, several steel grades have been developed in the past and are being manufactured industrially and used by the consumers worldwide. These steel grades, particularly for structural application, are categorized into conventional and different generations (1st, 2nd and 3rd generation) based on their strength-ductility combination that is the product of ultimate tensile strength (UTS) and total elongation (TE) [2]. This product of UTS and TE also indicates the energy absorption capability of the steel when subjected to tensile loading. Hence, an improvement in either the strength or ductility or both will improve the performance of the steel used in an application. This is because the higher strength will allow using a relatively thinner section, which will reduce the weight of a component and fuel consumption without compromising on safety. Moreover, steel with better ductility will prevent early failure of the component. However, for a majority of steel grades, the higher strength is usually accompanied by a fairly poor ductility, and vice-versa [3–9]. Therefore, in recent times, a significant amount of research has been focused on achieving improved strength-ductility combination. Some of the ways for a simultaneous improvement of both strength and ductility are increased alloying additions, use of novel processing routes or additional heat treatments, strategic alloy design, etc. [3–9]. The higher alloying addition in steel not only increases the cost but also leads to numerous issues during steel making and casting, its processing, fabrication and welding. Considering all these, two new low alloy steels have been designed, and processed through the water quenching route, to achieve a good combination of strength and ductility. The variation in mechanical properties has been discussed in the light of alloy composition and processing schedule.

16.2 Alloy Composition, Processing and Characterization

The chemical composition, carbon equivalent and M_s temperature of the studied alloys (i.e. Alloy-1 and Alloy-2) are mentioned in Table 16.1. The M_s temperature and the carbon equivalent (CE) were empirically estimated using Andrews equation [10] and Dearden and O'Neill formula [11], respectively. The vacuum induction furnace of 40 kg capacity was used to make the alloys in the form of a square ingot with size $10 \times 10 \times 50$ cm^3. This as cast ingot was cut into small pieces of $10 \times 10 \times 5$ cm^3, which was then homogenized at 1200 °C for 2 h and subsequently hot forged to make a plate of 8 mm thickness. This plate was then further cut into small pieces for homogenization at 1200 °C for 2 h, followed by hot rolling to reduce the thickness from 8 to 2.4 mm (total 70% reduction in thickness), and subsequently quenched in water to room temperature. To identify the phases present in the sample, Bruker D8 Advance X-ray diffractometer (XRD) with Cu-Kα radiation and Ni filter was used. The SU7000 Hitachi scanning electron microscope (SEM) 430 was used

Table 16.1 Chemical composition (wt%), empirical carbon equivalent (CE) and M_S temperature of the investigated alloys

Alloy	C	Mn	Si	Cr	Mo	Ni	V	Ti	B	Fe	CE	M_S (°C)
Alloy-1	0.16	1.1	0.5	0.2	0.1	0.5	0.03	–	–	Bal.	0.44	425.9
Alloy-2	0.21	1.47	0.5	0.2	0.1	0.55	–	0.025	0.001	Bal.	0.55	392.6

for microstructural analysis. The EMCO Test DuraScan Vickers hardness testing machine was used for hardness measurements. The flat tensile samples of gauge length and gauge width of 50 mm and 12.5 mm, respectively, were tested using universal tensile machine of 250 kN at a cross head velocity of 1 mm/min, as per ASTM E8M standard.

16.3 Results and Discussion

The results of XRD scan on Alloy-1 and 2 are shown in Fig. 16.1. Both the alloys show the presence of peaks corresponding to BCC/BCT type structure, indicating a major presence of ferrite and/or martensite. Additionally, small peak corresponding to FCC structure was observed in Alloy-1, which signifies the presence of small amount of retained austenite. However, the peak is too small for precise quantification, therefore, the amount of retained austenite in the Alloy-1 is expected to be below 5 vol%. To further confirm the presence of various phases and their distribution, scanning electron micrographs were also captured for both the samples and the results are shown in Fig. 16.2. The microstructure of Alloy-1 was observed to be dominated by the presence of martensite containing fine carbides within the martensite laths (Fig. 16.2a, b). Furthermore, a small amount of ferrite and thin-film type retained austenite was also present in the microstructure. This is in contrast to the microstructure of Alloy-2 (Fig. 16.2c, d), which only showed the presence of martensite laths containing fine carbides. These microstructural observations are consistent with the XRD scan results. The fine carbide precipitation in both the alloys could be due to self-tempering after martensite formation below M_S temperature, during the water quenching to room temperature. Similarly, a small amount of retained austenite has also been stabilized due to carbon partitioning from martensite to untransformed austenite. Both of these phenomena like carbon diffusion to austenite and carbide precipitation have happened, even during fast water quenching, due to the fact that carbon being interstitial atoms do not require a significantly higher temperature and prolonged holding for its movement. Between both the alloys, only Alloy-1 showed the presence of small amount of ferrite. This is due to a lower carbon and Mn content in the Alloy-1, in contrast to Alloy-2.

The mechanical properties of both the alloys are mentioned in Table 16.2. Alloy-2 showed higher hardness (about 65 H_V increase) in comparison to Alloy-1. In terms of tensile properties, both the alloys exhibited ultra-high strength (~1400–1600 MPa

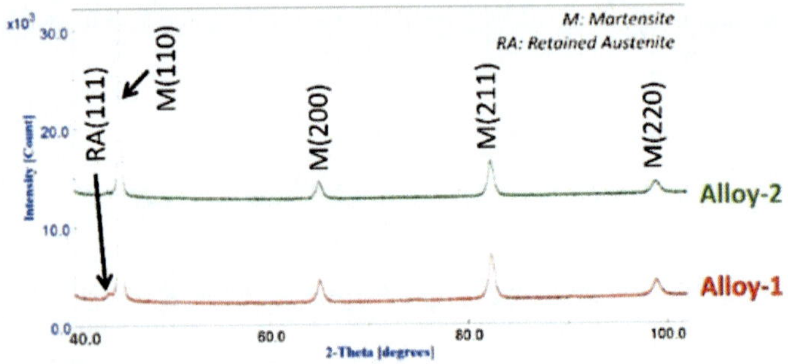

Fig. 16.1 XRD profiles of the investigated alloys

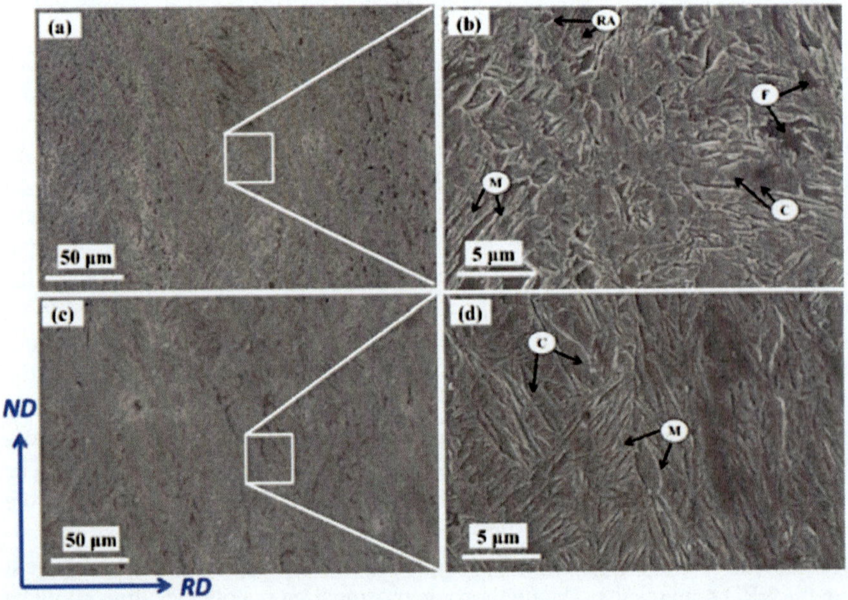

Fig. 16.2 Scanning electron micrographs of **a**, **b** alloy-1 and **c**, **d** alloy-2 (M: martensite, RA: retained austenite, C: carbides, F: ferrite)

UTS) combined with good ductility (~11–14% total elongation). As a result, the energy absorption capability, which is the product of ultimate tensile strength (UTS) and total elongation (TE), was also measured to be on the higher side (~18–20 GPa%). Similarly, Alloy-2 showed superior strength (about 100–150 MPa increase) with a slight compromise on the ductility (about 2% lower), when compared with Alloy-1. For example, Alloy-2 achieved the yield strength (YS) of 1056.5 ± 8.85 MPa, which was reduced to 962.5 ± 52.4 MPa for Alloy-1. Moreover, Alloy-2 could achieve the UTS of 1581.7 ± 12.2 MPa, which was reduced to 1447.7 ± 12.2 MPa for Alloy-1. The uniform elongation and total elongation for Alloy-1 was measured to

Table 16.2 Mechanical properties of Alloy 1 and 2

Mechanical properties	Alloy 1	Alloy 2
Hardness (H_V 1)	445 ± 11	510 ± 8
Yield strength, YS (MPa)	962.5 ± 52.4	1056.5 ± 8.85
Ultimate tensile strength, UTS (MPa)	1447.7 ± 12.2	1581.7 ± 12.2
Uniform elongation, UE (%)	8.05 ± 0.79	7.45 ± 0.16
Total elongation, TE (%)	13.19 ± 1.18	11.63 ± 0.38
YS/UTS ratio	0.66 ± 0.03	0.67 ± 0.01
UTS × TE (GPa%)	19.1 ± 1.7	18.4 ± 0.7

be 8.05 ± 0.79 and 13.19 ± 1.18%, respectively. These values of uniform elongation and total elongation were somewhat reduced to 7.45 ± 0.16 and 11.63 ± 0.38%, respectively for Alloy-2. This higher strength and reduced ductility for Alloy-2 could be attributed to its higher carbon and Mn content than Alloy-1, which has provided additional substitutional and interstitial strengthening of the martensite. Furthermore, the presence of small amount of ferrite and retained austenite has also contributed to reduced strength but higher ductility in Alloy-1.

16.4 Conclusion

The microstructural characterization and XRD investigation highlighted the presence of multiphase structure containing martensite, ferrite, retained austenite and carbides in the alloy with lower carbon and Mn content (Alloy-1). However, the ferrite formation was avoided in the alloy with higher carbon and Mn content (Alloy-2). The carbide precipitation and austenite retention at room temperature, even during fast water quenching, can be ascribed to the faster carbon diffusion due to its smaller size. A higher carbon and Mn content in the Alloy-2 has lead to higher hardness and strength with a slight decrease in the ductility, when compared with Alloy-1. The mechanical performance of both the alloys was found to be superior, i.e. ultra-high strength (~900–1100 MPa YS and ~1400–1600 MPa UTS) combined with sufficient ductility (~11–14% total elongation).

Acknowledgements This work was sponsored by Tata Steel Ltd., Jamshedpur. The authors acknowledge the Director, CSIR-NML and Tata Steel management for their kind encouragement and permission to publish this work. The authors are also thankful to IIEST Shibpur for providing their hot rolling and heat treatment facilities.

References

1. National Steel Policy (NSP).: https://steel.gov.in/national-steel-policy-nsp-2017 (2017)
2. AHSS Application Guidelines 6.0.: https://www.worldautosteel.org/downloads/599700/
3. Bansal, G.K., Srivastava, V.C., Chowdhury, S.G.: Role of solute Nb in altering phase transformations during continuous cooling of a low-carbon steel. Mater. Sci. Eng. A **767**, 138416 (2019)
4. Bansal, G.K., Madhukar, D.A., Chandan, A.K., Ashok, K., Mandal, G.K., Srivastava, V.C.: On the intercritical annealing parameters and ensuing mechanical properties of low-carbon medium-Mn steel. Mater. Sci. Eng. A **733**, 246–256 (2018)
5. Chandan, A.K., Bansal, G.K., Kundu. J., Chakraborty, J., Chowdhury, S.G.: Effect of prior austenite grain size on the evolution of microstructure and mechanical properties of an intercritically annealed medium manganese steel. Mater. Sci. Eng. A **768**, 138458 (2019)
6. Bansal, G.K., Rajinikanth, V., Ghosh, C., Srivastava, V.C., Kundu, S., Chowdhury, S.G.: Microstructure–property correlation in low-Si steel processed through quenching and nonisothermal partitioning. Metall. Mater. Trans. A **49**, 3501–3514 (2018)
7. Bansal, G.K., Junior, L.P., Ghosh, C., Rajinikanth, V., Tripathy, S., Srivastava, V.C., Bhagat, A.N., Chowdhury, S.G.: Quench temperature-dependent phase transformations during nonisothermal partitioning. Metall. Mater. Trans. A **51**, 3410–3424 (2020)
8. Bansal, G.K., Rajinikanth, V., Ghosh, C., Srivastava, V.C., Dutta, M., Chowdhury, S.G.: Effect of cooling rate on the evolution of microstructure and mechanical properties of nonisothermally partitioned steels. Mater. Sci. Eng. A **788**, 139614 (2020)
9. Bansal, G.K., Pradeep, M., Ghosh, C., Rajinikanth, V., Srivastava, V.C., Bhagat, A.N., Kundu, S.: Evolution of microstructure in a low-Si micro-alloyed steel processed through one-step quenching and partitioning. Metall. Mater. Trans. A **50**, 547–555 (2019)
10. Andrews, K.W.: Heat treatment for improvement in low temperature mechanical properties of 0.40 pct C–Cr steels. J. Iron Steel Inst. **203**, 721–727 (1965)
11. Dearden, J., O'Neill, H.: A guide to the selection and welding of low alloy structural steels. Trans. Inst. Weld. **3**, 203–214 (1940)

Chapter 17
Effect of Aging on Microstructure and Mechanical Behavior of Superalloy 617

Sajad Hamid and Ujjwal Prakash

Abstract The mechanical behaviour of superalloy 617 displayed a tremendous variation after certain hours of aging heat treatment from the as received material which was received in homogenized condition. The microstructural analysis of the as received specimen showed scarce presence of second phase particles and only some primary carbo-nitrides were present which are being thought to evolve during the solidification process of the alloy. Carbides mainly $M_{23}C_6$ evolved with the aging at temperature (750 °C) which brought a significant change in the microstructure. The nucleation of the Cr and Mo rich $M_{23}C_6$ takes place mainly on the grain boundaries and on the interfaces of Ti(C, N) particles. This formation of $Cr_{23}C_6$ in the grain interior renders the site rich in Ti and depleted in Cr and C. This provides the driving force for the nucleation of Ni_3(Al, Ti), an intermetallic compound exhibiting ordered FCC crystal structure, nucleates coherently with the austenitic matrix (γ). These help in improving the mechanical strength at room temperature as well as at elevated temperatures. Hardness values goes up with the aging treatment. Yield strength and tensile strength were also found to increase significantly. The already present carbo-nitrides provide interfaces for the nucleation of carbides and precipitates without coarsening and hence avoid microstructural degradation. These evolve finely and hence impede dislocation motion effectively. The increase in hardness values along with yield strength and tensile strength is accompanied by the decrease in ductility. Carbides were directly observed and analyzed in the aged specimen using microscopic techniques like optical microscopy and scanning electron microscopy. The carbide particles present along the grain boundaries prevent grain boundary sliding, resist plastic deformation at elevated temperatures. This makes them a suitable material and meets the requirements to withstand the exposures at elevated temperatures and hence are considered as the primary candidate for application in steam turbine power plants.

S. Hamid (✉) · U. Prakash
Indian Institute of Technology, (IITR), Roorkee, Uttarakhand 247667, India
e-mail: shamid@mt.iitr.ac.in

A. Tewari et al. (eds.), *Proceedings of the 3rd International Conference on Advances in Materials Processing: Challenges and Opportunities*, Springer Proceedings in Physics 293, https://doi.org/10.1007/978-981-99-1971-0_17

17.1 Introduction

Ni-based superalloys have been widely employed in aero engines and other modern turbine engines over the last several decades owing to their excellent high-temperature tensile, creep, and fatigue properties [1]. Alloy 617 belongs to the group of nickel-base alloys usually referred to as solid-solution strengthened materials [2]. Ni-based superalloys have been under extensive research because of their wide applications in industries at elevated temperatures. These also find applications in chemical laboratories and industries for their excellent resistance to corrosive atmospheres. Ni-based superalloys are primary candidate materials for structural applications at high temperatures owing to their high temperature microstructural stability. Superalloy 617 is a Ni-based superalloy which qualifies to be used for high temperature applications. This alloy exists as single austenitic phase at room temperature. This alloy is readily weldable. The conventional techniques are employed for welding these alloy components. Due to the ever-increasing demand for extending the life time of superalloys at high temperatures, a great amount of research effort has been made to improve their creep deformation resistance. Superalloy 617 is expected to withstand the stress of 100 MPa at 750 °C for 100,000 h exposure without enough creep deformation. The main objective of the proposed high-temperature usage of the IN617 alloy at the AUSC power plants is to increase power generation efficiency [2]. This alloy provides excellent resistance to high-temperature corrosion and oxidation.

Superalloy 617 is a solid-solution strengthened, nickel–chromium–cobalt–molybdenum alloy. The composition of this alloy has been tailored for better performance at high temperatures. Boron along with the little tolerances in other elements modified it to 617B. Addition of boron improves the weldability of the superalloy. Creep life also increased after the addition of Boron [3]. The higher weight percent of Chromium up to 23% present in the austenitic matrix, not only imparts solid solution strengthening but also provides excellent oxidation resistance by forming (Cr_2O_3), a very stable Oxide, during service exposure at higher temperatures [3]. The primary microstructural change in this superalloy is caused by formation of inter and intra granular carbides with a composition of $M_{23}C_6$ or M_6C during long-term exposure. Ni_3(Al, Ti), an intermediate solid solution formed by Ti, Al, and Ni, increases the creep strength. This mainly evolves at the interface between the matrix and the already-existing phase Ti(C, N) [4], at intermediate temperatures (550–850 °C). The microstructure of this alloy under exposure has an appreciable impact on its mechanical behaviour [2]. The mechanical properties such as yield strength, tensile strength and toughness show a significant difference.

Superalloy 617 CCA grade was made available and used for analysis in the current work. Inconel 617 is a solid solution strengthened Ni-based superalloy with a small amount of gamma prime (γ') present [5]. In addition to solid solution strengthening, the grain structure of Inconel 617 is stabilized by precipitation of carbides and some strengthening of the alloy is derived from carbides and carbo-nitrides of types $M_{23}C_6$ and Ti(C, N). Titanium and aluminium may additionally contribute to precipitation hardening through intermetallic precipitation of γ'. After exposure to high temperature 1000 °C and creep conditions, precipitation and coarsening continually remove the strengthening elements from the γ-matrix and significantly degrade the high temperature creep and mechanical strength of the alloy. Several structural characterization studies have been conducted on thermally and creep-exposed Inconel 617 alloys. These papers have been primarily focused on the precipitation of $M_{23}C_6$ and Ti(C, N) in the thermal and creep exposed alloy and there has been less emphasis on the role of Ni_3Al (γ') on the microstructure.

17.2 Experimental Procedure

The material was received from MIDHANI LTD. (Hyderabad, India) in the form of cylindrical bar, homogenized in the as received condition. The material was taken for further analysis using wire EDM (Electric Discharging Machine). The samples were sliced after cutting and were made ready for characterization. Polishing was done using emery papers with grit sizes varying from 100 to 2000. The samples were loaded for aging treatment into the muffle furnace, its temperature was set at 750 °C. The samples were loaded in the block form and the aged samples were taken out after 100 h. The tensile specimen with 25 mm gauge length were taken from the as received samples as well as aged sample for characterization and mechanical testing. The samples for characterization and mechanical testing were cut and polished to mirror finish using emery papers of varying grit size ranging from 320 to 3000. The phase identification of the as received sample was done using X-ray Diffraction Analysis, Rigaku Automated X-Ray Diffractometer. Cu target was used as the X-ray source. Optical microscopy was done using Leica, Model: 5000 M optical microscope. Scanning Electron Microscopy was done for further characterization using Carl Zeiss Ultra Plus Scanning Electron Microscope. For Optical and Scanning Electron microscopy analysis. Aqua regia was used to etch the as received sample and for the aged sample Kaling's reagent was used as the etchant.

17.3 Results and Discussion

17.3.1 Microstructural Characterization

The chemical composition of this alloy was analyzed by using Electron Probe Microanalysis Make: Cameca, the facility being provided by Institute Instrumentation center, IIT Roorkee (Table 17.1).

The elemental composition was averaged from five different locations along the polished surface of the specimen. Chromium is the major strengthening element

Table 17.1 Chemical composition of superalloy 617

Element	Al	Si	Ti	Cr	Mn	Fe	Co	Ni	Cu	Mo
Wt (%)	1.15	0.06	0.45	22.5	0.15	0.2	12.5	55	0.1	7.8

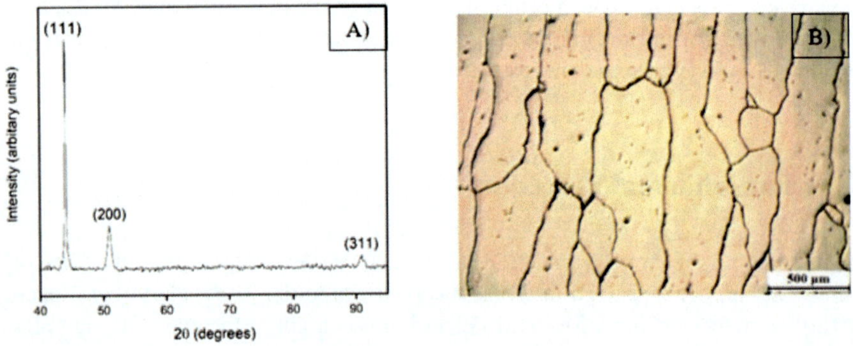

Fig. 17.1 Characterization of the as received sample **a** X-ray diffraction analysis of as-received sample. **b** Optical micrograph of the as received sample

followed by cobalt and molybdenum which are higher melting point elements and hence provide strengthening at higher temperatures.

The initial characterization was done using X-ray diffraction analysis and with microscopic techniques like Optical Microscopy (Fig. 17.1).

The aging treatment was done for 100 h of time and the mechanical behaviour displayed was appreciably different for aged specimen than as received one.

The higher magnification optical micrograph clearly shows the presence of particles around the triple junctions on grain boundaries and inside the grains with different shapes and sizes. Scanning electron microscopy also revealed the formation of carbides decorating the grain boundaries which is further examined by EDS in SEM (Fig. 17.2).

17.3.2 Mechanical Behaviour

Mechanical behaviour of both the samples was revealed using Hardness Testing machine and Tensile testing at room as well as at higher temperature (750 °C). The hardness was calculated by testing the specimen under Vickers's microhardness testing machine at load 10 kgf and applied for 20 s. Hardness of the aged sample was appreciably higher than the as received sample. The other mechanical properties that showed the difference are the increase in yield strength and ultimate tensile strength for the aged sample as compared to the as received sample (Fig. 17.3). The improvement in mechanical strength is accompanied by the decrease in ductility for

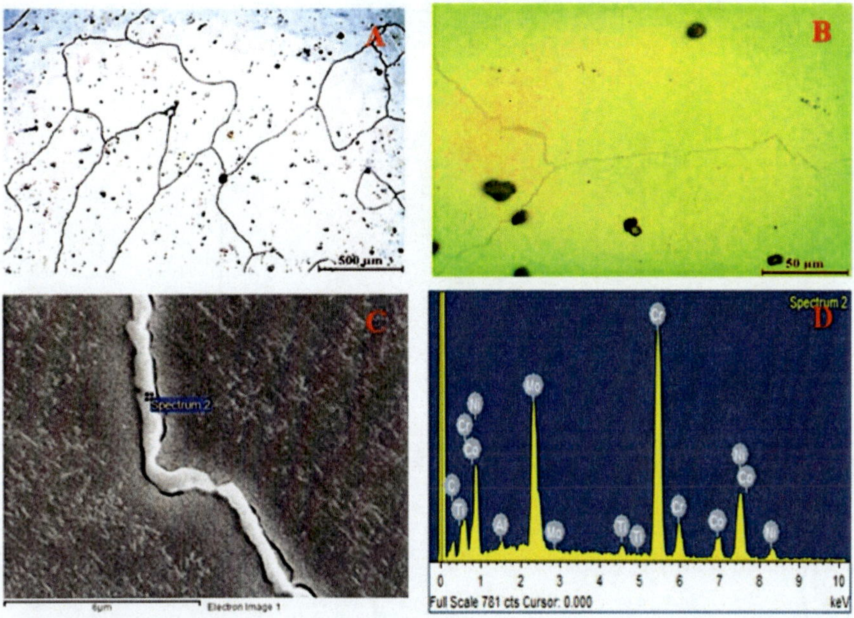

Fig. 17.2 Characterization of aged sample: **a** optical micrograph of the aged sample. **b** Optical micrograph of the aged sample at higher magnification **c** secondary electron image of grain boundary region using SEM. **d** EDS analysis shows the site rich in Cr and Mo

the aged samples. The increased yield strength clearly shows that resistance to the plastic deformation is due to the impediment of dislocation movement by the presence of these carbide and precipitate particles. The nucleation and growth of these second phase particles responsible for strengthening the austenitic matrix depend upon the exposure temperature and time.

The characterization techniques make the presence of precipitates and carbides evident in the aged specimen. The presence of these precipitates and carbide particles caused a significant impact on the mechanical properties of the alloy. The hardness value along with yield strength and tensile strength increases after aging treatment (Fig. 17.3). The ductility was drastically reduced after aging which confirmed further the formation of secondary phase particles that hindered the movement of dislocations as well as grain boundaries at elevated temperatures.

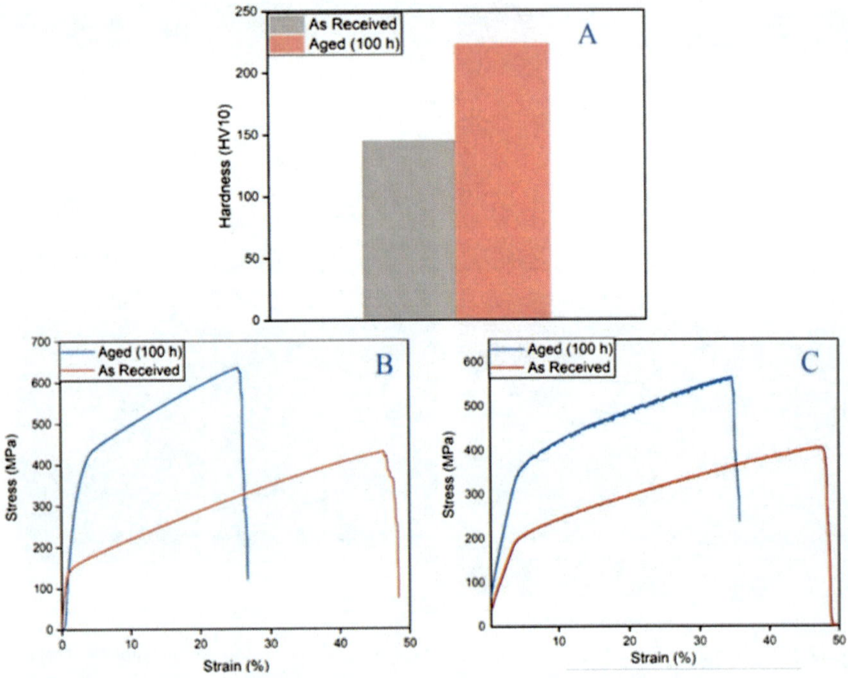

Fig. 17.3 Mechanical testing. **a** Hardness testing of the samples using Vickers's microhardness measurement for as received and aged samples. **b** Tensile test data at room temperature. **c** Tensile test data at high temperature (750 °C)

17.4 Conclusions

1. Microscopy techniques clearly demonstrate the presence of carbides; EDS (Energy Dispersive Spectroscopy) analysis reveals that Cr and Mo are enriched preferentially at grain boundary sites. The enrichment is caused by the formation of carbides near these sites. During the ageing process, these precipitates and carbides also develop inside the grains, and these play an important role in deciding the mechanical behaviour of this alloy both at room temperature and at higher temperatures.
2. Ti(C, N) is a primary carbide is present initially after solidification, provides interface for the evolution of $Cr_{23}C_6$ inside the grains.
3. Carbides forming inside the grains are beneficial than those forming on the grain boundaries because the rate of diffusion is slow in the grain interior and hence the rate of microstructural degradation caused by the dissolution and coarsening of carbides is not too much.
4. The hardness and yield strength enhances after aging, which is due to the resistance to the dislocation motion during plastic deformation which is provided by precipitates and carbides.

These precipitates and carbides enhance the strength at higher temperatures which is due to the pinning of grain boundaries by these carbide particles that prevents grain boundary sliding.

Acknowledgements I am grateful to my Supervisor, Professor Ujjwal Prakash for providing me all the support technically and creating a pathway for my research work. I am thankful to IIT Roorkee, especially to the Department of Metallurgical and Materials Engineering for creating a platform to work in the pleasant environment and providing me facilities for carrying out all the experiments. I would like to be grateful to MHRD for providing the funds for my research work.

References

1. Sims and Hagel, 1972; Sims, 1984; Pollock and Tin, 2006
2. Guo, Y., Wang, B., Hou, S.: Aging precipitation behavior and mechanical properties of Inconel 617 superalloy. Acta Metall. Sin. (English Lett.) **26**, 307–312 (2013). https://doi.org/10.1007/s40195-012-0249-3
3. The Minerals, Metals & Materials Society and ASM International (2020). https://doi.org/10.1007/s11661-020-06066-8
4. Wu, Q., et al.: Microstructure of Long-Term Aged IN617 Ni-Base Superalloy
5. Krishna, R., Atkinson, H.V., Hainsworth, S.V., Gill, S.P.: Gamma prime precipitation dislocation densities, and TiN in creep-exposed Inconel 617 alloy. Metall. Mater. Trans. A Phys. Metall. Mater. Sci. **47**, 178–193 (2016). https://doi.org/10.1007/s11661-015-3193-9

Chapter 18
Derivation of Unified Interaction Parameter Formalism and Its Application in Local Equilibria Demarcation of Fe–C–X (X = Mn, Si and Cr) Systems

Vikash Kumar Sahu, Snehashish Tripathy, Sandip Ghosh Chowdhury, and Gopi Kishor Mandal

Abstract In the present work, the thermodynamically consistent Unified Interaction Parameter Formalism (UIPF) proposed by Pelton and Bale has been derived from the interaction parameter formalism proposed by Lupis and Elliot. In addition, the UIPF has been adopted for the estimation of ferrite–austenite and cementite–austenite phase boundaries at 500 °C in Fe–C–X systems, where X corresponds to a substitutional component (X = Mn, Si, Cr). Furthermore, local equilibria demarcated phase boundaries at 500 °C in Fe–C–X systems have also been presented, which can be helpful in determining phase transformation kinetics.

Keywords Unified interaction parameter formalism · Local equilibria · Phase transformation

18.1 Introduction

The advent of computational tools for the evaluation of phase diagrams and other thermodynamic properties have contributed immensely in designing newer steels with distinctive microstructures and consequent properties. However, most of the industrially relevant high strength steels are processed at temperatures where ortho-equilibrium (OE) conditions usually do not prevail. It has rather been shown that other metastable local equilibria (LE) models such as Negligible Partitioning Local Equilibrium (NPLE), Partitioning Local Equilibrium (PLE) and Paraequilibrium (PE), explain the phase transformation behavior more accurately than OE [1–3]. The application of these LE models in unveiling the phase transformation behavior has led

V. K. Sahu (✉) · S. Tripathy · S. G. Chowdhury · G. K. Mandal
Academy of Scientific and Innovative Research (AcSIR), Ghaziabad 201002, India
e-mail: Vikash.nml19a@acsir.res.in

CSIR-National Metallurgical Laboratory (CSIR-NML), Jamshedpur 831007, India

A. Tewari et al. (eds.), *Proceedings of the 3rd International Conference on Advances in Materials Processing: Challenges and Opportunities*, Springer Proceedings in Physics 293, https://doi.org/10.1007/978-981-99-1971-0_18

to the design of optimized alloy compositions and process parameters for various advanced high strength steel grades in the recent past [2]. However, for designing more and more industrially feasible low alloyed high strength steels, the knowledge of metastable solvus curves along with various limiting LE conditions at low temperature ranges are required, the availability of which, has been one of the major concerns of steel researchers globally.

The solvus curves and the various limiting LE conditions can be evaluated by equating the partial molar free energy of a component i in phase ϕ which is given by Eq. (18.1). The terms $G_i^{0\phi}$, R, T and a_i^{ϕ} represent standard Gibbs energy of component i in phase ϕ, universal gas constant, absolute temperature and activity of component i in phase ϕ. Equation (18.1) can be further written as given by Eq. (18.2), where γ_i^{ϕ} is the activity coefficient of component i in phase ϕ. The $RT \ln \gamma_i^{\phi}$ term in Eq. (18.2) is the partial molar excess free energy of component i which represents the non-ideal interactions between the components in phase, ϕ. Various available thermodynamic formalisms adopt different expressions of $\ln \gamma_i^{\phi}$ for accurate estimation of excess free energy contribution while taking the interactions between various components into consideration.

$$\overline{G_i^{\phi}} = G_i^{0\phi} + RT \ln a_i^{\phi} \tag{18.1}$$

$$\overline{G_i^{\phi}} = G_i^{0\phi} + RT \ln x_i^{\phi} + RT \ln \gamma_i^{\phi} \tag{18.2}$$

18.2 Some Important Thermodynamic Formalisms

In some of the studies pertaining to advanced high strength steel design, researchers have adopted Wagner Interaction Parameter Formalism (WIPF) for estimation of excess free energy owing to its computational simplicity [4, 5]. However, WIPF is thermodynamically inconsistent at finite solute concentration (as it doesn't satisfy Gibbs Duhem equation), and therefore the obtained thermodynamic interaction parameters may not be reliable [6, 7]. A thermodynamically consistent Unified Interaction Parameter Formalism (UIPF) was proposed by Pelton and Bale given by Eqs. (18.3), (18.4) and (18.5). The objective was to make the Wagner Interaction Parameter Formalism thermodynamically consistent at all concentrations. In this paper, 1 refers to solvent and 2, 3 refer to solutes.

$$\ln \gamma_1 = -\frac{1}{2}\varepsilon_2^2 x_2^2 - \varepsilon_2^3 x_2 x_3 - \frac{1}{2}\varepsilon_3^3 x_3^2 \tag{18.3}$$

$$\ln \gamma_2 = \ln \gamma_2^0 + \varepsilon_2^2 x_2 + \varepsilon_2^3 x_3 - \frac{1}{2}\varepsilon_2^2 x_2^2 - \varepsilon_2^3 x_2 x_3 - \frac{1}{2}\varepsilon_3^3 x_3^2 \tag{18.4}$$

$$\ln \gamma_3 = \ln \gamma_3^0 + \varepsilon_2^3 x_2 + \varepsilon_3^3 x_3 - \frac{1}{2}\varepsilon_2^2 x_2^2 - \varepsilon_2^3 x_2 x_3 - \frac{1}{2}\varepsilon_3^3 x_3^2 \tag{18.5}$$

18.2.1 Equivalence of UIPF with the Formalism by Lupis and Elliot

Lupis and Elliot had proposed the expression for $\ln \gamma_i$ about the point of infinite dilution as given in Eq. (18.6) [8].

$$\ln \gamma_i = J_{0,0}^{(i)} + J_{1,0}^{(i)} x_2 + J_{0,1}^{(i)} x_3 + J_{2,0}^{(i)} x_2^2 + J_{1,1}^{(i)} x_2 x_3 + J_{0,2}^{(i)} x_3^2 + \ldots \quad \text{(where } i = 1, 2, 3) \tag{18.6}$$

Since the solvent obeys the Raoultian standard state, the coefficient $J_{0,0}^{(1)}$ in the expression of $\ln \gamma_i$ for solvent equals to zero $\left((\ln \gamma_1)_{x_1 \to 1} = 0\right)$. The terms $J_{0,0}^{(2)}$ and $J_{0,0}^{(3)}$ in the expression of $\ln \gamma_i$ for solutes (2 and 3) represent the $\ln \gamma_2$ and $\ln \gamma_3$ value at infinite dilution of all solutes, given by $\ln \gamma_2^0$ and $\ln \gamma_3^0$ respectively (with respect to Raoultian standard state). The coefficients $J_{1,0}^{(i)}$, $J_{0,1}^{(i)}$, $J_{1,1}^{(i)}$, $J_{2,0}^{(i)}$ and $J_{0,2}^{(i)}$ can be expressed in the form of partial differential equation as shown in Eq. (18.7).

$$J_{1,0}^{(i)} = \left(\frac{\partial \ln \gamma_i}{\partial x_2}\right)_{x_1 \to 1}, \quad J_{0,1}^{(i)} = \left(\frac{\partial \ln \gamma_i}{\partial x_3}\right)_{x_1 \to 1}, \quad J_{1,1}^{(i)} = \left(\frac{\partial^2 \ln \gamma_i}{\partial x_2 \partial x_3}\right)_{x_1 \to 1},$$
$$J_{2,0}^{(i)} = \left(\frac{\partial^2 \ln \gamma_i}{\partial x_2^2}\right)_{x_1 \to 1} \quad \text{and} \quad J_{0,2}^{(i)} = \left(\frac{\partial^2 \ln \gamma_i}{\partial x_3^2}\right)_{x_1 \to 1} \tag{18.7}$$

Substituting the coefficients $J_{i,j}^k$ with the interaction parameters (ε_2^2, ε_2^3, ε_3^2, ε_3^3 ρ_2^2, $\rho_2^{2,3}$, ρ_2^3, ρ_3^2, ρ_3^3 and $\rho_3^{2,3}$) in Eq. (18.6) for solutes (2 and 3) results into the new expressions given by Eqs. (18.8) and (18.9).

$$\ln \gamma_2 = \ln \gamma_2^0 + \varepsilon_2^2 x_2 + \varepsilon_2^3 x_3 + \rho_2^2 x_2^2 + \rho_2^{2,3} x_2 x_3 + \rho_2^3 x_3^2 + \ldots \tag{18.8}$$

$$\ln \gamma_3 = \ln \gamma_3^0 + \varepsilon_3^3 x_3 + \varepsilon_3^2 x_2 + \rho_3^2 x_2^2 + \rho_3^{2,3} x_2 x_3 + \rho_3^3 x_3^2 + \ldots \tag{18.9}$$

Upon substituting the expression of $\ln \gamma_1$, $\ln \gamma_2$ and $\ln \gamma_3$ from Eqs. (18.6), (18.8) and (18.9) in the Gibbs Duhem relation given by Eq. (18.10), it results in the expression given by Eqs. (18.11) and (18.12).

$$(1 - x_2 - x_3)\frac{\partial \overline{G}_1^{ex}}{\partial x_i} + x_2 \frac{\partial \overline{G}_2^{ex}}{\partial x_i} + x_3 \frac{\partial \overline{G}_3^{ex}}{\partial x_i} = 0 \quad \text{(where } i = 2, 3) \tag{18.10}$$

$$(1 - x_2 - x_3)\left(J_{1,0}^{(1)} + 2J_{2,0}^{(1)}x_2 + J_{1,1}^{(1)}x_3 + \ldots\right)$$
$$+ x_2\left(\varepsilon_2^2 + 2\rho_2^2 x_2 + \rho_2^{2,3}x_3 + \ldots\right)$$
$$+ x_3\left(\varepsilon_3^2 + 2\rho_3^2 x_2 + \rho_3^{2,3}x_3 + \ldots\right) = 0 \quad (18.11)$$

$$(1 - x_2 - x_3)\left(J_{0,1}^{(1)} + 2J_{0,2}^{(1)}x_3 + J_{1,1}^{(1)}x_2 + \ldots\right)$$
$$+x_2\left(\varepsilon_2^3 + \rho_2^{2,3}x_2 + 2\rho_2^3 x_3 + \ldots\right)$$
$$+x_3\left(\varepsilon_3^3 + \rho_3^{2,3}x_2 + 2\rho_3^3 + x_3 + \ldots\right) = 0 \quad (18.12)$$

Upon further simplification, Eqs. (18.11) and (18.12) can be re-written in the form of expression given by Eq. (18.13).

$$A + Bx_2 + Cx_3 + Dx_2^2 \ldots = 0 \quad (18.13)$$

Equation (18.13) can only be zero if all the coefficients (*A*, *B* and *C* etc.) are equal to zero, which ultimately yield the relations given by Eq. (18.14).

$$J_{1,0}^{(1)} = 0 = J_{0,1}^{(1)},\ J_{2,0}^{(1)} = -\frac{1}{2}\varepsilon_2^2,\ J_{1,1}^{(1)} = -\varepsilon_2^3 = -\varepsilon_3^2 \text{ and } J_{0,2}^{(1)} = -\frac{1}{2}\varepsilon_3^3 \quad (18.14)$$

Upon substituting the values of $J_{0,0}^{(1)}$, $J_{1,0}^{(1)}$, $J_{2,0}^{(1)}$, $J_{1,1}^{(1)}$ and $J_{0,2}^{(1)}$ in Eq. (18.6), the expression for $\ln \gamma_1$ can be written as shown in Eq. (18.15).

$$\ln \gamma_1 = -\frac{1}{2}\varepsilon_2^2 x_2^2 - \varepsilon_2^3 x_2 x_3 - \frac{1}{2}\varepsilon_3^3 x_3^2 + \ldots \quad (18.15)$$

Equation (18.16) represents the integral molar excess Gibbs energy expression, which can further be written as given by Eq. (18.17) by substituting the expression for $\ln \gamma_1$, $\ln \gamma_2$ and $\ln \gamma_3$.

$$\frac{G_m^{xs}}{RT} = (1 - x_2 - x_3)\ln \gamma_1 + x_2 \ln \gamma_2 + x_3 \ln \gamma_3 \quad (18.16)$$

$$\frac{G_m^{xs}}{RT} = x_2 \ln \gamma_2^0 + x_3 \ln \gamma_3^0 + \frac{1}{2}\varepsilon_2^2 x_2^2 + \frac{1}{2}\varepsilon_3^3 x_3^2 + \varepsilon_2^3 x_2 x_3 \ldots \quad (18.17)$$

Relationship between partial molar excess Gibbs energy of each component and the integral molar excess Gibbs energy in a ternary system can be represented by Eq. (18.18). On substituting the second order truncated expression of integral molar excess Gibbs energy from Eq. (18.17) in (18.18), expression for $\ln \gamma_1$, $\ln \gamma_2$ and $\ln \gamma_3$ can be evaluated. The obtained expressions for $\ln \gamma_1$, $\ln \gamma_2$ and $\ln \gamma_3$ are identical with the ones proposed by Pelton and Bale for UIPF given by Eqs. (18.3), (18.4) and (18.5). Since the expression for $\ln \gamma_1$, $\ln \gamma_2$ and $\ln \gamma_3$ have been derived using the Gibbs Duhem relation, the formalism thus obtained for $\ln \gamma_i$ is thermodynamically consistent.

$$\overline{G}_i^{xs} = RT \ln \gamma_i = G_m^{xs} - x_2 \frac{\partial G^{xs}}{\partial x_2} - x_3 \frac{\partial G^{xs}}{\partial x_3} \tag{18.18}$$

18.3 Metastable Phase Boundaries Prediction Using UIPF

The equality of partial molar excess free energy of component i between two phases given by Eq. (18.19) represents a tie line in the two phase region (i.e. ferrite–austenite, cementite–austenite and ferrite–cementite). Using the expression of $\ln \gamma_i$ obtained from the UIPF, Eq. (18.19) can be expanded and written in terms of interaction parameters and standard Gibbs energy. The interaction parameters and standard Gibbs energy change values have been evaluated using the equilibrium composition across tie lines obtained from ThermoCalc® (TCFE 9 database) at high temperatures, where the ThermoCalc® data are considered to be highly accurate. These obtained interaction parameters and the standard Gibbs energy change have been linearly extrapolated to lower temperatures, which have been used in the evaluation of phase boundaries at stable and metastable temperatures. For demonstration purpose, the ferrite/austenite (α/γ) and cementite/austenite (θ/γ) phase boundaries evaluated at 500 °C for Fe–C–Mn, Fe–C–Cr and Fe–C–Si systems using UIPF have been shown in Fig. 18.1. The UIPF predicted phase boundaries have been found to be in good agreement with ones simulated from ThermoCalc®. From Fig. 18.1a it can be seen that for Fe–C–Mn system, ThermoCalc® can predict ferrite austenite phase boundaries only in certain composition range, whereas UIPF can predict the phase boundaries for full composition range. Similarly, for Fe–C–Cr system at 500 °C, ThermoCalc® could not predict ferrite austenite solvus curves which can be easily predicted using UIPF.

$$\overline{G}_i^{\text{phase1}} = \overline{G}_i^{\text{phase2}} \tag{18.19}$$

18.4 PLE/NPLE Boundary Prediction Using UIPF

Partitioning local equilibrium and Negligible partitioning local equilibrium demarcation in two phase field helps in positioning whether the composition is lying in PLE or NPLE envelope. Knowing the position of composition helps in determining the thermodynamic properties and the kinetics considering the respective local equilibrium

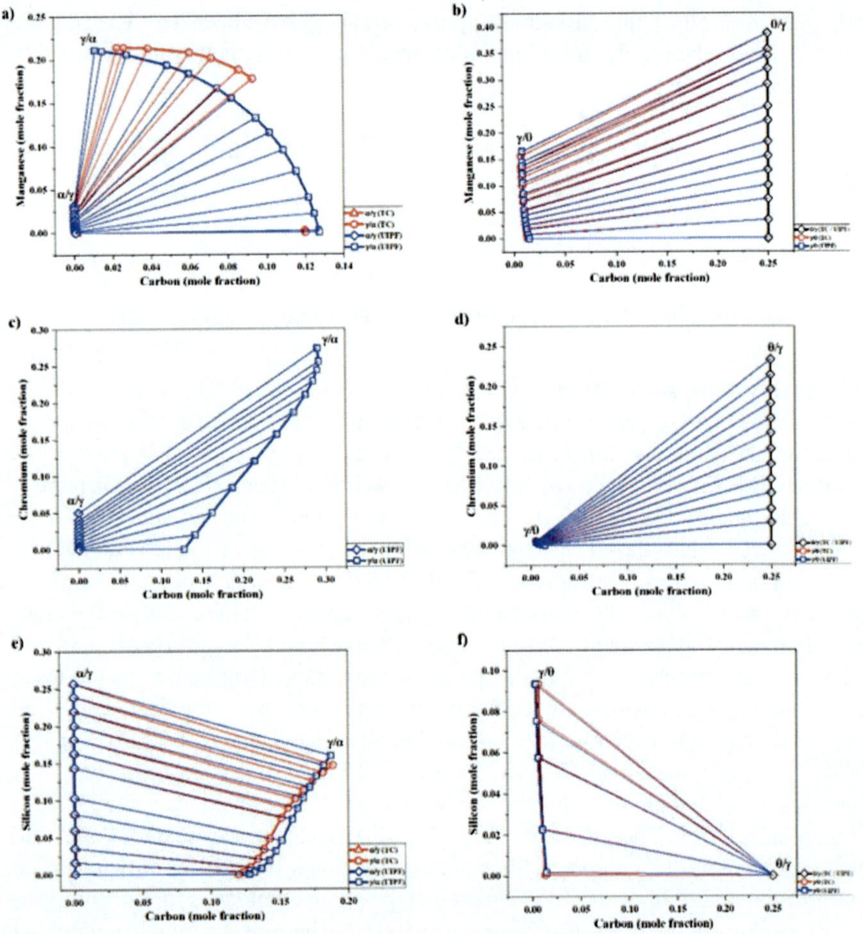

Fig. 18.1 UIPF and ThermoCalc® predicted phase boundaries at 500 °C for **a** α/γ phase field in Fe–C–Mn, **b** θ/γ phase field in Fe–C–Mn, **c** α/γ phase field in Fe–C–Cr, **d** θ/γ phase field in Fe–C–Cr, **e** α/γ phase field in Fe–C–Si and **f** θ/γ phase field in Fe–C–Si systems

modes [1–3]. NPLE/PLE demarcated ferrite/austenite and cementite/austenite phase boundaries using UIPF at 500 °C for Fe–C–Mn, Fe–C–Cr and Fe–C–Si systems have been shown in Fig. 18.2.

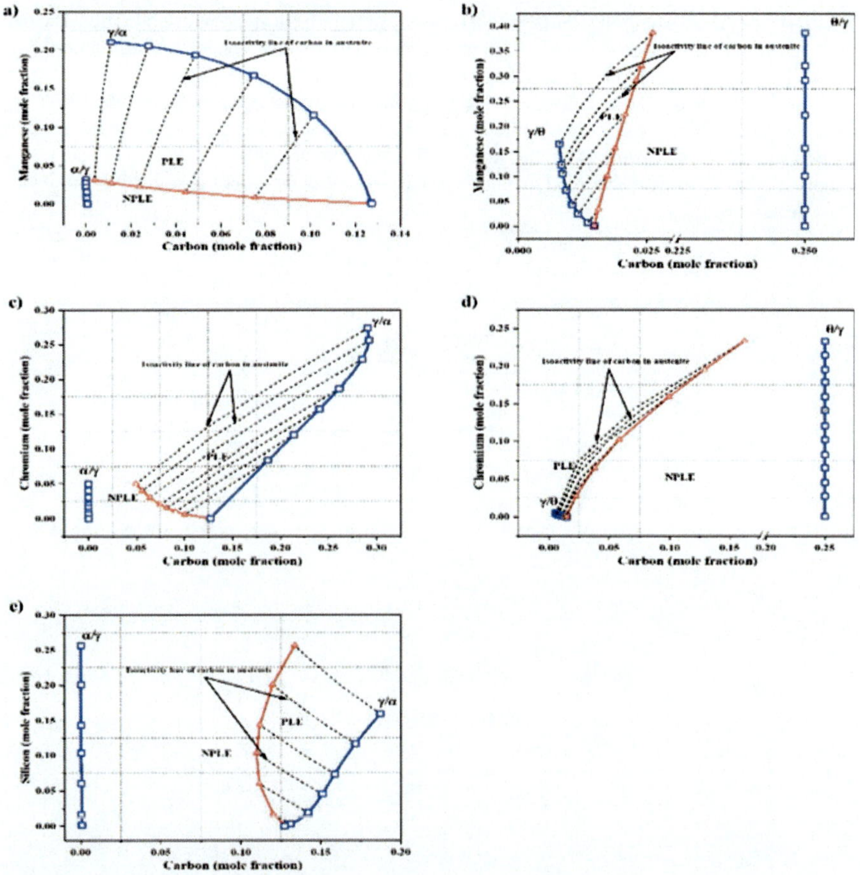

Fig. 18.2 NPLE/PLE demarcated phase diagrams using UIPF at 500 °C in **a** α/γ phase field of Fe–C–Mn; **b** θ/γ phase field of Fe–C–Mn; **c** α/γ phase field of Fe–C–Cr, **d** θ/γ phase field of Fe–C–Cr, and **e** α/γ phase field of Fe–C–Si

References

1. Tripathy, S., Jena, P.S.M., Sahu, V.K., Sarkar, S.K., Ahlawat, S., Biswas, A., Mahato, B., Tarafder, S., Chowdhury, S.G.: On the competitive substitutional partitioning during nano pearlitic transformation in multicomponent steels. Metall. Mater. Trans. A **53**, 1806–1820 (2022)
2. Tripathy, S., Sahu, V.K., Jena, P.S.M., Tarafder, S., Chowdhury, S.G.: On the importance of local equilibria in alloy design criteria for bulk nano-pearlitic steels and ensuing mechanical properties. Mater. Sci. Eng., A **841**, 143034 (2022)
3. Tripathy, S., Sahu, V.K., Chowdhury, S.G., Mandal, G.K.: On the unified interaction parameter formalism and its application in critical reassessment of pearlitic transformation in Fe–C–Mn system. Manuscript under consideration in Metall. Mater. Trans. A
4. Dai, Z., Ding, R., Yang, Z., Zhang, C., Chen, H.: Elucidating the effect of Mn partitioning on interface migration and carbon partitioning during quenching and partitioning of the Fe–C–Mn–Si steels: modeling and experiments. Acta Mater. **144**, 666–678 (2018)

5. Fang, H., van der Zwaag, S., van Dijk, N.H.: A novel 3D mixed-mode multigrain model with efficient implementation of solute drag applied to austenite–ferrite phase transformations in Fe–C–Mn alloys. Acta Mater. **212**, 116897 (2021)
6. Darken, L.: Thermodynamics of ternary metallic solutions. Trans. AIME **239**, 90–96 (1967)
7. Bale, C.W., Pelton, A.D.: The unified interaction parameter formalism: thermodynamic consistency and applications. Metall. Trans. A **21**, 1997–2002 (1990)
8. Lupis, C.H.P.: Chemical Thermodynamics of Materials. Elsevier Science Publishing Co. Inc., New York (1983)

Chapter 19
Erosive Wear of Low Temperature Nitrided 13/4 MSS for Hydro-turbine Application

Akeshwar Singh Yadav, Nitin Kumar, Gaurav Mahendru, Guru Prakash, and S. K. Nath

Abstract Wear and fatigue are the serious concerns for engineering structures and components like turbine blades, crank shaft, thermal power plant, gears etc. An investigation on wear behavior of nitrided 13/4 martensitic stainless steel resulting from low temperature salt bath nitriding is presented in this work. 13/4 martensitic stainless steel was nitrided in salt bath at 500 °C. The nitrided layer thickness, hardness and erosion wear behavior of the treated material is studied. Results show that low temperature treatment produced a hard layer without forming chromium nitride precipitates, which is the greatest challenge during nitriding of stainless steels, thus keeping stainless properties of the steels intact. The slurry erosion behavior of nitrided steels shows that 4 hour nitriding resulted in minimum erosion.

Keywords 13–4 MSS · Slurry erosion · Hydro turbine · Salt bath nitriding

19.1 Introduction

Hydraulic turbine parts are made of 13Cr–4Ni martensitic stainless steel (13/4 MSS) due to its excellent mechanical and corrosion resistant properties [1, 2]. However, wear of the hydraulic components mainly runner blades, impeller, guide vanes, and nozzles etc., are the main problem among them. When the fluid carrying solid

A. S. Yadav (✉)
Jindal Steel and Power Ltd., Raigarh, CG 496001, India
e-mail: akeshwarsinghyadav@gmail.com; akeshwar.yadav@jindalsteel.com

A. S. Yadav · N. Kumar · G. Prakash · S. K. Nath
Indian Institute of Technology Roorkee, Roorkee 247667, India
e-mail: nkumar@mt.iitr.ac.in

G. Mahendru
Applied Materials India Pvt. Ltd., Whitefield Road, Bangalore 560066, India

G. Prakash
Metallurgical Engineering Department, O P Jindal University, Raigarh 496109, India

A. Tewari et al. (eds.), *Proceedings of the 3rd International Conference on Advances in Materials Processing: Challenges and Opportunities*, Springer Proceedings in Physics 293, https://doi.org/10.1007/978-981-99-1971-0_19

particles strike the surface of material, it erodes its surface by various mechanisms depending on the surface properties. This process is named as slurry erosion. These erodents (silt particles) consist of hard quartz (70–98%) in major portion and cause damages like leakage, efficiency falls, and drop in structure strength [3]. Wear starts from the surface plastic deformation due to the impact of erodent particles. To prevent erosion, various techniques can be employed like coating the substrate by physical/chemical vapor deposition, microstructure modification, modifying the chemical composition at the surface of substrate (by nitriding, carburizing) etc.

Nitriding is a widely used method to enhance the tribological properties of steels [4]. Nitriding causes diffusion of nitrogen at the material surface. This N led to the generation of high surface compressive stresses and offers higher wear resistance together with the corrosion resistance properties [4–6]. Nitriding can be done by any of the methods viz. gas, salt bath or plasma. Salt bath nitriding is comparatively much faster than gas nitriding and is economical than plasma nitriding. Relatively high temperature nitriding (above 510 °C) gives high hardness and faster N diffusion kinetics but due to the formation of CrN, corrosion resistance properties are lost significantly [7, 8]. Existing low temperature nitriding literature suggest that after nitriding one can get precipitation free steel with supersaturated nitrogen [4, 7–10].

The present work utilizes the cost effective salt bath nitriding treatment to enhance the surface properties of widely used 13/4 MSS without disturbing its core properties. In this study, slurry erosion behaviour of nitrided 13/4 MSS is investigated.

19.2 Experimental Details

The composition of 13/4 MSS checked by using the optical emission spectroscopy (Thermo Jarrell Ash) is reported in Table 19.1.

Samples with dimension 10 mm × 15 mm × 25 mm were used in this work. These samples were polished up to 1500 grit size SiC papers followed by ultrasonic cleaning in acetone. KOCN based salt bath was used for nitriding [9]. The salt bath nitriding was done for various times viz. 1, 2, 4, 6 and 8 h in a muffle furnace at 500 °C, followed by the water quenching. Further examinations of the nitriding layer were carried out by scanning electron microscope and micro hardness indenter. Slurry erosion test was done on the bare and nitrided surfaces. Sand particles with hardness range between 900 and 1100 HV used in the slurry erosion test are sieved by the sieve tester. In this test, sand particles of 210–320 μm were used. Slurry of 1 kg sand (10 wt%) and 9 L distilled water (90 wt%) were prepared for the pot tester. The relative velocity between the samples and slurry was set at 3.18 m/s. Tests were

Table 19.1 Composition analysis of 13/4 MSS

Element	C	Si	Mn	P	S	Cr	Ni	Cu	Mo	Fe
Wt%	0.01	0.51	0.46	0.004	0.01	14.2	3.03	0.13	0.37	Rest

done for 24 h time period, but a periodic mass loss measurement was done after each 3 h cycle.

19.3 Results and Discussion

19.3.1 Nitriding Treatment

Salt bath nitriding treatments were performed on the steel samples of fully martensitic microstructure (Fig. 19.1a) for various times. It has been observed that the thickness of nitrided layer increases with time. This incremental thickness of the nitrided material (for selected conditions) presented in Fig. 19.1 is taken at the sample cross-section near the surface area. After 1 h nitriding, an average layer of 3.5 μm is obtained. Nitriding is a diffusional process. So, as the treatment time increases, layer thickness also increases. Nitriding layers for 8 h have a layer thickness of 29.3 μm. Figure 19.2a shows the variation of nitriding thickness after different time treatments.

Hardness measurements were also conducted along the cross-section of the nitrided samples from the surface to inside. The hardness variation across the cross-section is presented in Fig. 19.2b. The hardness of the nitriding layer increases with

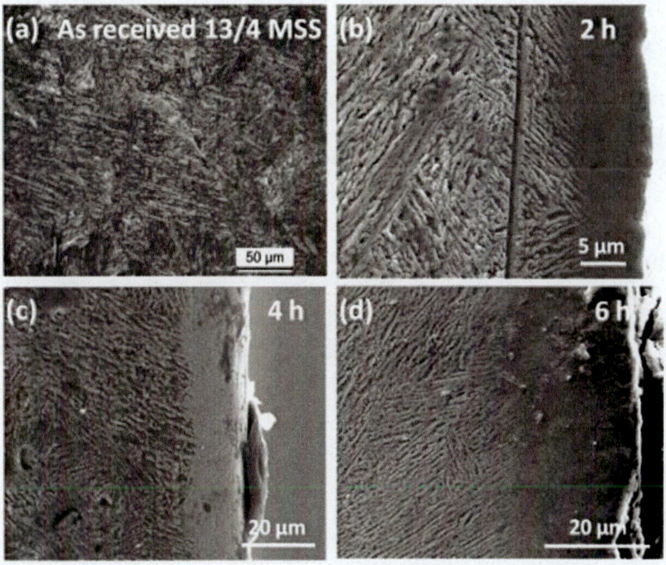

Fig. 19.1 **a** As-received sample with fully martensitic microstructure. Nitrided layer at the surface cross-section (at right side in microstructure, unetched/shiny region) after the treatment at **b** 2 h, **c** 4 h and **d** 6 h

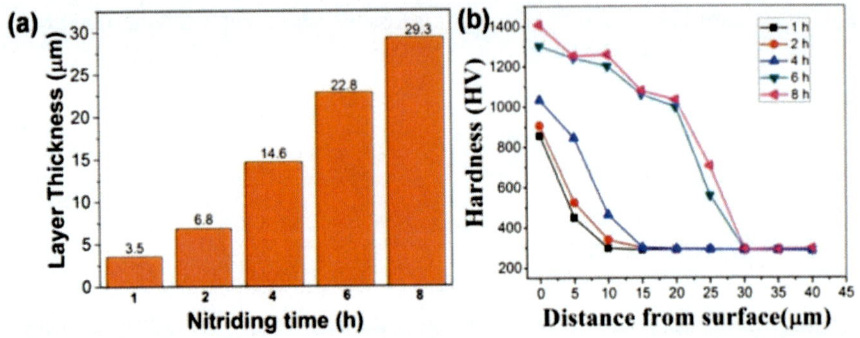

Fig. 19.2 **a** Average nitrided layer thickness for samples nitrided for various times. **b** Variation of surface hardness for different nitrided samples with respect to distance from the surface

the increase in nitriding time. Maximum hardness value was observed in case of 8 h nitriding.

19.3.2 Erosive Wear of Nitrided MSS

Slurry erosion results give a deep insight on the effect of nitriding. Weight loss is presented in Fig. 19.3e. Maximum wear loss is observed for the as received sample. Initially the rate of weight loss is very high but later it proceeds towards the steady state condition. The minimum weight loss observed for 4 h nitrided sample.

SEM image of slurry eroded samples were taken to study the wear mechanism. The surface of the as received sample shows the presence of micro cutting, plastic flow, craters and pits (Fig. 19.3a). In 2 h nitrided sample (Fig. 19.3b), a lesser material erosion is observed compared to the as received sample as indicated by the presence of less number of craters. For 4 h nitrided sample (Fig. 19.3c), the minimum material flow mark with less number of shallow craters indicates the further low surface erosion. In case of 6 h nitrided sample (Fig. 19.3d), surface morphology together with the presence of micro crack indicates the on-set of brittle kind of wear behavior.

19.3.3 Discussion

The N content at the surface increases with the treatment time and thus, increases the nitriding layer thickness (Fig. 19.2a) and also increases the residual compressive stress at the surface. At lesser time nitriding (i.e. 1, 2 and 4 h), only the expanded martensite is formed, which holds the nitrogen in the solid solution. Hardness value is directly proportional to the amount of nitrogen diffused which further depends on the time of nitriding (Fig. 19.2b). In samples nitrided for longer times (6 and 8 h), surface

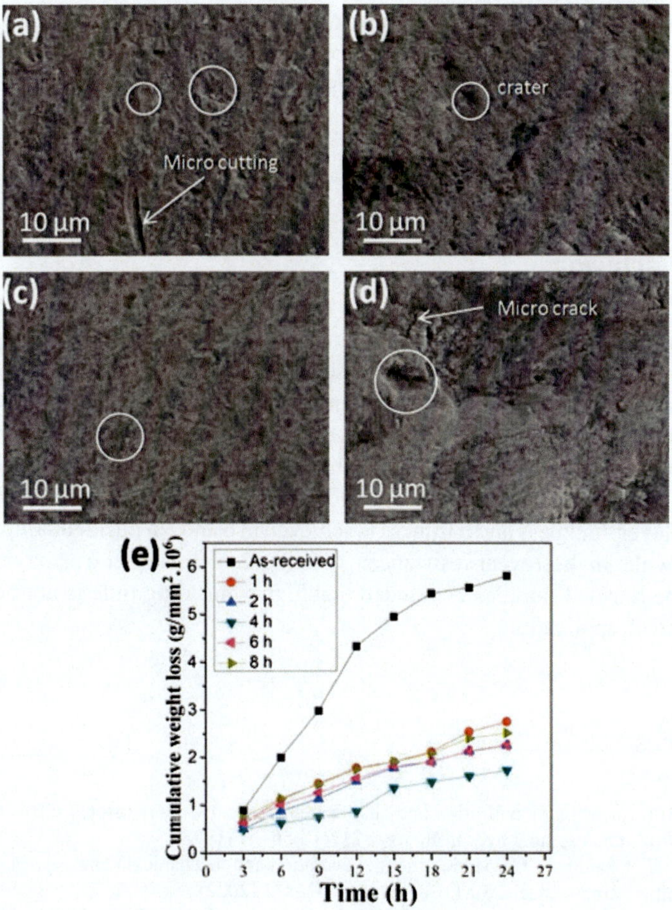

Fig. 19.3 SEM micrographs after slurry erosion of **a** as received, **b** 2 h nitrided, **c** 4 h nitrided, **d** 6 h nitrided samples. **e** Cumulative weight loss curve for different nitrided samples

hardness is not only enhanced by N supersaturation but also by the precipitation of chromium nitride (CrN). This phenomenon in turn degrades the corrosion resistance properties of the steel. The core microstructure of the MSS material is also affected by the heat treatment due to tempering of martensite. This can led to improve the toughness of the core which makes the material to sustain higher water pressure/loads during the operations. Further investigations are required to verify this.

Sample nitrided for 4 h shows the highest erosion wear resistance. Wear microstructures suggest the transformation of ductile to brittle kind of material removal (Fig. 19.3) when nitriding was performed for longer times (above 4 h). Research shows that ductile materials have maximum material removal when erodent particle strikes it at 30° and for brittle materials the maximum material removal occurs at the particle incident angle of 90°. Wear of ductile and brittle material takes place by cutting mechanism and brittle fracture, respectively. Hardness of the as received material is least which means it has lesser ability to resist erosion. But, in 8 h nitrided sample, even though the hardness is highest, it does not show the best wear resistance. This indicates that surface hardness is not the only criteria which determines the rate of material removal.

19.4 Conclusion

To minimize the slurry erosion of the hydraulic machinery, a simple low temperature salt bath nitriding process is used in the present work. 13Cr–4Ni martensitic stainless steel was subjected to nitriding for different intervals of time at 500 °C. Optimum nitriding time was found to be the 4 h, which shows a reduction of about 70% in weight loss due to erosion wear as compared to the un-nitrided sample. A transition from ductile to brittle material removal is observed with increased nitriding time. Although, maximum layer thickness and hardness is achieved in 6 and 8 h nitrided samples, they do not show the highest wear resistance. Also, the depletion of corrosion resistance in long time nitrided samples concluded that higher nitriding time is not beneficial for this kind of application.

References

1. Mann, B.S.: Erosion visualisation and characteristics of a two dimensional diffusion treated martensitic stainless steel hydrofoil. Wear **217**(1), 56–61 (1998)
2. Mohan, N., Chaudhari, G.: Microstructural evolution and mechanical behaviour of boronized martensitic stainless steels. Surf. Eng. **38**(5), 552–561 (2022)
3. Tong, D.: Cavitation and wear on hydraulic machines. Int. WP DC Int. **3**, 37–44 (1981)
4. Mittemeijer, E.J.: Fundamentals of nitriding and nitrocarburizing, ASM Handbook. In: Steel Heat Treating Fundamentals and Processes, vol. 4A, pp. 619–646 (2013)
5. Li, C.X., Bell, T.:Corrosion properties of plasma nitrided AISI 410 martensitic stainless steel in 3.5% NaCl and 1% HCl aqueous solutions. Corros. Sci. **48**, 2036–2049 (2006)
6. Sun, Y., Bell, T., Wood, G.: Wear behaviour of plasma-nitrided martensitic stainless steel. Wear **178**, 131–138 (1994)
7. Alphonsa, I., Chainani, A., Raole, P.M., Ganguli, B., John, P.I.: A study of martensitic stainless steel AISI 420 modified using plasma nitriding. Surf. Coat. Technol. **150**, 263–268 (2002)
8. Li, C.X., Bell, T.: Corrosion properties of active screen plasma nitrided 316 austenitic stainless steel. Corros. Sci. **46**(6), 1527–1547 (2004)
9. Kumar, N., Chaudhari, G.P., Meka, S.R.: Investigation of low-temperature liquid nitriding conditions for 316 stainless steel for improved mechanical and corrosion response. Trans. Indian Inst. Met. (2019)
10. Prakash, G., Nath, S.K.: Studies on enhancement of silt erosion resistance of 13/4 martensitic stainless steel by low-temperature salt bath nitriding. J. Mater. Eng. Perform (2018)

Chapter 20
Ultrasonic Shot Peening and Its Influence on Oxidation Behavior of 347 Grade Austenitic Stainless Steel

Amit Kumar Gupta, Ghanshyam Das, and Kausik Chattopadhyay

Abstract The newly developed austenitic stainless-steel grade 347 is widely used in thermal power plants as a thick steam pipe component for heat exchanger, steam super-heater and re-heater applications. This alloy has requisite mechanical properties like creep performance and oxidation resistance under steam and furnace exhaust gases over an operating temperature range of 500–800 °C. Although, these steels also suffer from oxidation, oxide scale exfoliation and hot corrosion after long time exposure to severe environment or at higher temperature. Adhesion of oxide scale is a crucial factor to know the oxidation resistance of an alloy. Ultrasonic shot peening (USSP) is a prominent process for surface treatment to enhance mechanical properties and corrosion resistance of a material. The main objective of this research work is to study the mechanism of high temperature oxidation in air environment and its improvement through surface modification by USSP. USSP was done on SS347 for different time periods of 0.0, 2.0 and 3.0 min. There was improvement in the oxidation resistance at 750 °C for SS347 by USSP treatment at different time intervals.

Keywords 347 stainless steel · Ultrasonic shot peening · Air oxidation

20.1 Introduction

Corrosion and Oxidation at high temperature are one of the foremost issues in materials in various types of applications of re-heater, superheater and heat exchanger tubes in steam and coal power plants. These pipes are vulnerable to severe oxidative

A. K. Gupta (✉) · G. Das
Department of Materials and Metallurgical Engineering, National Institute of Advanced Manufacturing Technology, Hatia, Ranchi 834003, India
e-mail: pmme_042021@niamt.ac.in

K. Chattopadhyay
Department of Metallurgical Engineering, Indian Institute of Technology, Banaras Hindu University (IIT-BHU), Varanasi 221005, India

A. Tewari et al. (eds.), *Proceedings of the 3rd International Conference on Advances in Materials Processing: Challenges and Opportunities*, Springer Proceedings in Physics 293, https://doi.org/10.1007/978-981-99-1971-0_20

and corrosive medium that contains ash with sulphur and alkali metal oxides [1]. Use of non-conventional fuel is best to preserve the environments but the availability and use of conventional fuels (such as coal, oil and gas) cannot be ignored. Coal is abundant and easy to reach and inexpensive in much of the area of the globe. Hence to produce the cost-effective electricity the coal fired steam power plants are best in business [2, 3]. Thermal efficiency of steam power plants can be enhanced by increasing the temperature of steam which enters into the turbine from 580 to 620 °C giving a thermal efficiency rise from 38 to 42% [4, 5]. For high working temperature, material selection plays an important role for the required equipment, to examine the corrosion resistance and life span of stainless steels to amend the material selection [6]. Austenitic stainless steel is best suitable for high temperature applications and are frequently used where corrosion resistance is much concerned [7]. Heat resisting austenitic stainless steels are the high temperature alloys mainly containing chromium and Nickel; e.g. 18% Cr and 8% Ni are referred to as grade 304 which is the basic composition of the austenitic stainless steel [8, 9]. Austenitic stainless steels (SS) of grades 321, 347 and 348 are the stabilized steels having outstanding resistance against intergranular corrosion which are exposed to temperatures in the Cr-carbides precipitation range from 427 to 816 °C [10]. These stabilized SS of grades 321 and 347 have excellent resistance to sensitization and creep failure. Therefore, these are used frequently for high temperature equipment, e.g., chemical reactors, boilers, heat exchanger tubes and nuclear reactors [11–14]. These alloys contain the elements like Cr, Si, Ni and carbon which affect the generation of passivating oxide growth, and due to this reason corrosion resistance of the alloy increases. Cr has a crucial role for the growth of the protective oxide layer on the surface as chromium oxide (Cr_2O_3) [15–17]. When Nb is added in type 347SS, NbC precipitates which stabilizes the grain boundary against the precipitation of $M_{23}C_6$ types of carbides because it consumes the chromium from grain boundary and open on to sensitization [13, 18, 19]. Exfoliation of steam side oxide scale occurs on the superheaters and re-heaters tubes used in modern supercritical boilers running with steam temperature range of 580–680 °C [17]. Advancement in surface nanocrystallization technique is achieved by applying deformation rate with or without the increased temperature such as cold working, surface mechanical attrition treatment (SMAT), ultrasonic shot peening (USSP), warm forging and rolling [20–22]. USSP is a cold working surface treatment process consisting of small spherical balls, by applying the kinetic energy to the shots. It bombards on the surface with the help of a jet of compressed air, acts as a hammer creating small indentation imparting increased mechanical properties, strength, hardness and fatigue resistance to the workpiece which increases the lifetime of materials by increasing the resistance against oxidation, corrosion and fluctuating loads [23–26].

The objective of this work is to investigate the effect of USSP with different time periods on this alloy system to increase the oxidation resistance and to reduce oxide exfoliation by increasing the mechanical properties. This work is focused on simulating the air oxidation environments of SS347 at high temperature of 750 °C till 70 h and characterization of formed oxides and present phases was done by Optical microscopy, XRD, SEM and EDS analysis.

Table 20.1 Chemical composition of the as received SS347 obtained from the OES (wt%)

C	Si	Mn	S	P	Cr	Mo	Ni	Nb	Ta	Fe
0.065	0.468	1.46	<0.005	<0.003	18.4	0.201	10.3	0.589	0.421	Bal.

20.2 Materials and Methods

All the experiments in this study were carried out using the austenitic stainless steel SS347 grade procured in the form of super-heater tube. Rectangular shaped specimen of 10 mm × 10 mm × 5 mm thickness were prepared from the tube part and mechanically polished on 1000, 1500 and 2000 grit emery paper before USSP. Samples were cleaned with acetone before USSP to remove the dirt. Table 20.1 presents the chemical compositions of the samples which were evaluated using optical emission spectrometer (OES).

The USSP was performed on Stress Voyager (SONATS) machine with hardened steel balls of 3 mm diameter at frequency of 20 kHz for different durations of 2.0 and 3.0 min. The un-shot peened specimens are named as Un-USSP and the samples subjected to USSP for 2.0 and 3.0 min were named as USSP2.0 and USSP3.0 respectively. Electric resistance-heating tube furnace was used at 750 °C for the oxidation tests in air environment on the UnUSSP, USSP2.0 and USSP3.0 samples, placed separately in silica crucible. Oxidation tests were performed for different time durations of 2 h for the first 5 cycles, 4 h for the next 3 cycles, then 8 h for the next 3 cycles and 12 h for the last two cycles, so for a total duration of 70 h, in between cycles samples were removed and cooled for 15 min. Exfoliated oxide scale can be seen on the surface of the samples. Total weight gain is recorded including that of the crucible with the help of digital balance. Microstructural changes, oxide layers and formed oxides was examined with the help of optical microscopy (LEICA DM1750M). Aqua regia (10 mL of conc. HNO_3, 30 mL conc. HCl) etchant was used to get the optical micrographs. XRD (18 kW Rigaku miniflex 600 X-ray diffractometer using Cu-Kα radiation), SEM (ZEISS 840A operating at 30 kV) and EDS were done as the characterization techniques.

20.3 Results and Discussions

The optical micrograph (Fig. 20.1a) of the SS347 steel having austenite grains with annealing twin bands in several grains, the mean average length of grains is generally 35–45 μm [1, 26]. Figure 20.1b presents the optical micrograph of USSP2.0 samples which shows the dimple like structure formed by USSP. Size and shape of the dimples decreases with USSP time duration which are non-uniform and very much different from each other.

Oxidation test was done on 347SS in air exposed at 750 °C for a time duration of 70 h. The variation of weight gains per unit area (ΔW) mg/cm^2 with exposure

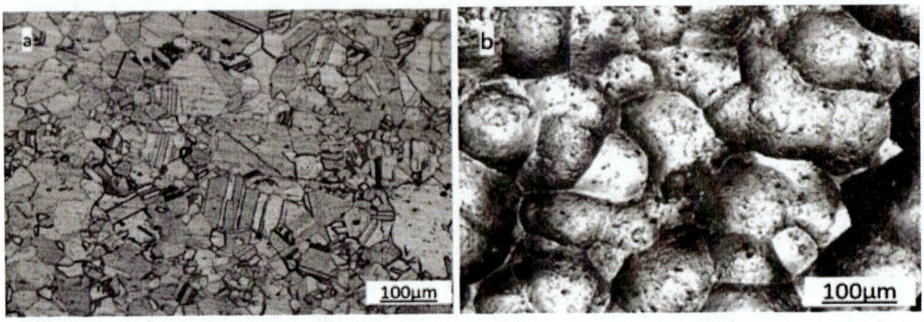

Fig. 20.1 Optical micrograph of 347SS **a** UnUSSP and **b** USSP2.0 samples

time (t) in hours are shown in Fig. 20.2. With the duration of oxidation, there is increase in the weight gain in oxidizing materials. By applying the equation: $(\Delta W)^n = K.t$, oxidation rate is derived mathematically where K is a rate constant in $mg^n\ cm^{-2n}\ h^{-1}$ and t is exposure time in hours, also $n.\ \ln(\Delta W) = \ln(K) + \ln(t)$ is derived from it and the linear plot of ΔW versus t was drawn in the Fig. 20.2. The values of n, K and R^2 were obtained from the oxidation kinetics and listed in Table 20.2. Here n values are approximately near 2, so assuming parabolic oxidation kinetics where K_p represents the parabolic rate constant. Earlier literature reported that most alloys of iron at high temperature in the range of 500–700 °C shows the parabolic oxidation kinetics [3–5, 26].

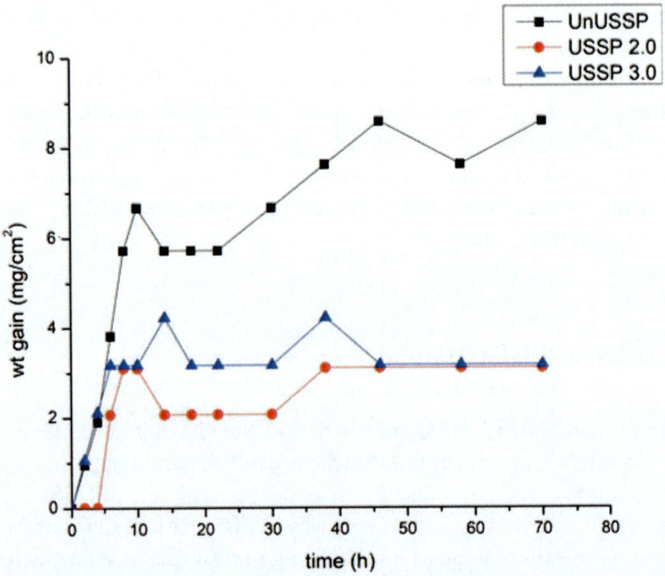

Fig. 20.2 Variation of weight gain with exposure time of the different UnUSSP and USSP treated samples of 347SS oxidized in air at 750 °C for 70 h

Table 20.2 Data of rate constant of oxidation of UnUSSP, USSP2.0 and USSP3.0 samples of 347SS oxidized in air at 750 °C for 70 h

	$(\Delta W)^n = k_p.t$		K_P $(mg^2\ cm^{-4}\ h^{-1})$	R^2 $(n = 2)$
Sample	n	R^2		
UnUSSP	1.85	0.72	0.0226	0.98
USSP 2.0	1.64	0.67	0.0422	0.92
USSP 3.0	1.53	0.62	0.0518	0.88

As shown in Fig. 20.3a, b, oxide scales formed on the surface were very thin. Peaks of the as received material Fe–C can be seen as γ-Fe and due to USSP the intensity of these peaks gets higher. Phase analysis of oxidized surface products of these Fe–Ni–Cr based austenitic steels along with C, H and O components contains spinel of Fe_2O_3, Fe_3O_4, $FeCr_2O_4$, Cr_2O_3, Ni_2O_3 and $NiCr_2O_4$. Cr_2O_3 forms a dense and adherent scale on the surface that provides obstacle for oxygen diffusion and protects the material [7, 13, 17, 26].

In order to characterize the oxide layer, microstructural changes and surface morphologies of the oxide films products formed on the specimens. SEM and EDS

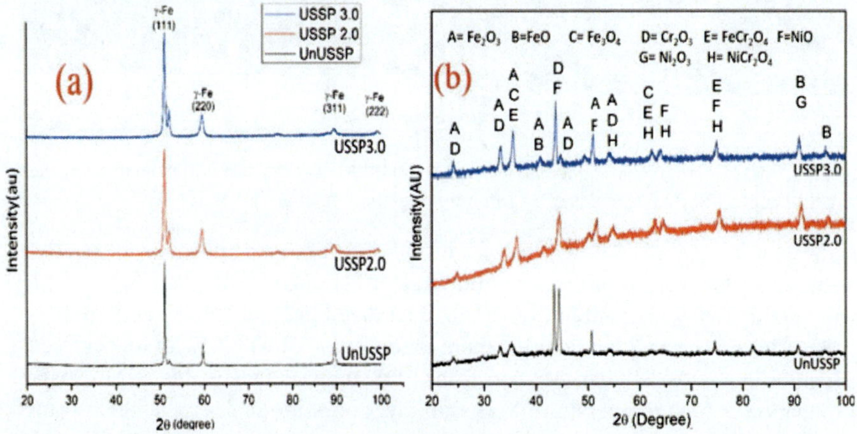

Fig. 20.3 XRD patterns of the Un-USSP and USSP-treated samples of 347SS **a** before and **b** after oxidation in air at 750 °C for 70 h

Table 20.3 Data of EDS quantitative analysis obtained from Fig. **b** UnUSSP and **d** USSP2.0 samples of 347SS after oxidation in air at 750 °C for 70 h

Elements (wt%)	C	O	Cr	Mn	Fe	Ni	Mo	Total
UnUSSP (Fig. b)	8.91	43.94	2.57	–	30.87	12.57	1.13	100.00
USSP2.0 (Fig. d)	–	30.96	20.06	5.77	41.11	1.53	0.57	100.00

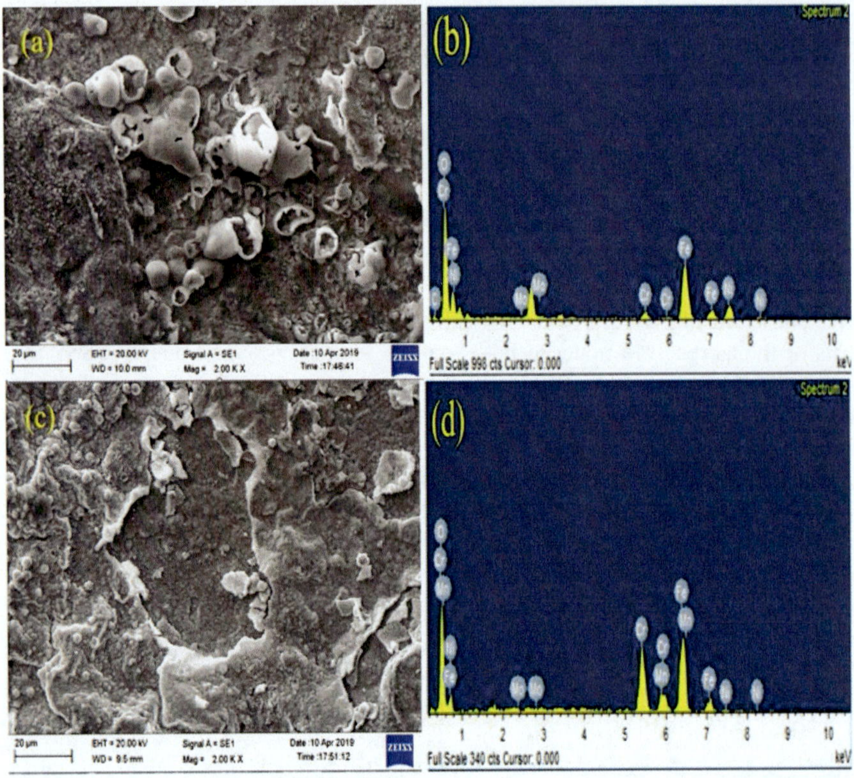

Fig. 20.4 SEM micrographs of the plan view of **a** UnUSSP, **c** USSP2.0 and EDS analysis of the **b** Un-USSP and **c** USSP2.0 samples of 347SS after oxidation in air at 750 °C for 70 h

analysis was done on the surface of stainless steel 347 of UnUSSP and USSP samples expressed in Table 20.3 and in Fig. 20.4a–d oxidized in air at 750 °C until 70 h.

From here, we can conclude that there is formation of Cr_2O_3 with low carbides. As carbon is less which can be seen from EDS analysis makes this steel resistant to oxidation at high temperature. Very thin layer of oxide scales have been formed on the sample which is confirmed by these micrographs and cannot be observed by optical microscope. The surface of oxidized 347SS reveals that the protective oxide layer at this high temperature range, is formed over the surface which contains Fe, O, Cr, Mo and Ni confirmed by EDS analysis and XRD pattern (Fig. 20.3b). In the USSP samples, there is a formation of much Cr_2O_3 than UnUSSP samples. Cr_2O_3 passivates the surface and forms a well adherent scale which enhances the oxidation resistance [1, 13, 26].

20.4 Conclusions

In this study, Correlation between oxidation behavior and surface modification treatment by ultrasonic shot peening of 347SS was investigated in air at 750 °C until 70 h. This steel contains 18 wt% chromium with the presence of niobium forms Cr_2O_3 and NbC, is good enough to enhance and give a better high temperature corrosion resistance than other steels.

Shot peening is a cold working surface treatment process imparts increased strength and high temperature oxidation resistance. Oxidation resistance at high temperature increases by increasing the time duration of shot peening. USSP3.0 is much resistant than USSP2.0 which can be seen from the parabolic oxidation rate (Fig. 20.2 and Table 20.2).

There is no phase transformation because of shot peening which is evident from XRD patterns. But as the shot peening time increases, the intensity of the peaks enhances. Formation of oxide scale with C, H and O components, as evident from SEM and quantitative EDS analysis, having oxides like Fe_2O_3, Fe_3O_4 and Cr_2O_3 forms on the surface of specimens. Cr_2O_3 acts as a barrier for oxygen diffusion into the matrix of specimens and provides well adherent scale on the surface that protects the material and makes it corrosion resistance at high temperature. Thus, oxidation resistance of USSP samples at high temperature like 750 °C, is more resistant than without shot peened sample due to increased mechanical properties and formation of protective oxides layer over the surface.

References

1. Viitala, H., Galfi, I., Taskinen, P.: Initial oxidation behaviour of niobium stabilized TP347H austenitic stainless steel—effect of grain size and temperature. (9), 5–7 (2015)
2. Fu, J., Zhou, Q., Li, N., Liu, Z., Liu, T.: Effects of external stresses on hot corrosion behavior of stainless steel. **104**, 103–111 (2016)
3. Li, Y., et al.: Early oxidation mechanism of austenitic stainless steel TP347H in supercritical water. Corros. Sci. **128**, 241–252 (2017)
4. Simms, T.D.M.ŁN., Sem, E.D.X.Á., Coal, Á.: Analysis of high temperature steam oxidation of superheater steels used in coal fired boilers. Oxid. Met. **85**(1), 171–187 (2016)
5. Ennis, P.J.Ã., Quadakkers, W.J.: Mechanisms of steam oxidation in high strength martensitic steels. **84**, 75–81 (2007)
6. Gomes, A., Navas, M., Uranga, N., Paiva, T., Figueira, I., Diamantino, T.C.: High-temperature corrosion performance of austenitic stainless steels type AISI 316L and AISI 321H, in molten Solar Salt. Sol. Energy **177**, 408–419 (2019)
7. Ghalambaz, M., Abdollahi, M., Eslami, A., Bahrami, A.: Case studies in engineering failure analysis short communication. A case study on failure of AISI 347H stabilized stainless steel pipe in a petrochemical plant. Case Stud. Eng. Fail. Anal. **9**, 52–62 (2017)
8. Malaysiana, S., et al.: Effect of high temperature corrosion on austenitic stainless-steel grade 304 in CO_2 gas at 700 °C. **43**(7), 1069–1075 (2014)
9. Covert, R.A., Tuthill, A.H.: Stainless steels: an introduction to their metallurgy and corrosion resistance. **20**(7), 506–517 (2000)
10. Chandra, K., Kain, V., Tewari, R.: Microstructural and electrochemical characterization of heat-treated 347 stainless steel with different phases. Corros. Sci. **67**, 118–129 (2013)

11. Tiamiyu, A.A., Eduok, U., Szpunar, J.A., Odeshi, A.G.: Corrosion behavior of metastable AISI 321 austenitic stainless steel: investigating the effect of grain size and prior plastic deformation on its degradation pattern in saline media. Sci. Rep. (2018), 1–18 (2019)
12. Mahajan, S., Chhibber, R.: Hot corrosion studies of boiler steels exposed to different molten salt mixtures at 950 °C. Eng. Fail. Anal. **99**(2018), 210–224 (2019)
13. Behnamian, Y., et al.: A comparative study of oxide scales grown on stainless steel and nickel-based superalloys in ultra-high temperature supercritical water at 800 °C. Eval. Program Plann. **106**(2017), 188–207 (2016)
14. Yu, J., Liu, S., Li, F., Wang, T.: Surface and coatings technology Na_2SiO_3/Al_2O_3 composite coatings on 304 stainless steels for enhanced high temperature oxidation inhibition and chloride-induced corrosion resistance. SCT **309**, 1089–1098 (2017)
15. Jinlong, L., Hongyun, L., Tongxiang, L.: Influence of pre-deformation and oxidation in high temperature water on corrosion resistance of type 304 stainless steel. J. Nucl. Mater. **466**, 154–161 (2015)
16. Lima, A.S., Nascimento, A.M., De Engenharia, D., De Engenharia, D.: Sensitization evaluation of the austenitic stainless steel AISI 304L, 316L, 321 and 347. 139–144 (2005)
17. Paper, C., Stefan, A., Oak, S., National, R.: Prediction of oxide scale exfoliation in steam tubes (2016)
18. Olumide, T., Lu, J.: Nano materials science literature review on the mechanical properties of materials after surface mechanical attrition treatment (SMAT). Nano Mater. Sci. (2020)
19. Kumar, V., Kumar Gupta, R., Das, G.: Influence of deformation and annealing on microstructure and corrosion behavior of austenitic stainless steel. Mater. Today, Proc. **22**, 3347–3352 (2019)
20. International, R.: Shot Peening to Prevent the Corrosion Cracking of Austenitic Stainless Steels (1979)
21. Wang, T., Yu, J., Dong, B.: Surface nanocrystallization induced by shot peening and its effect on corrosion resistance of $1Cr_{18}Ni_9Ti$ stainless steel. **200**, 4777–4781 (2006)
22. Kumar, S., Chattopadhyay, K., Mahobia, G.S., Singh, V.: Hot corrosion behaviour of Ti–6Al–4V modified by ultrasonic shot peening. JMADE **110**, 196–206 (2016)
23. Review, P.: From Portable Systems to Robotic Production Lines Ultrasonic Shot Peening (2016)
24. Pradhan, D., Mahobia, G.S., Chattopadhyay, K., Singh, V.: Effect of surface roughness on corrosion behavior of the superalloy IN718 in simulated marine environment. J. Alloys Compd. **740**, 250–263 (2018)
25. Peltz, S., Vanessa, L., Beltrami, R., Raquel, S.: Effect of the shot peening process on the corrosion and oxidation resistance of AISI430 stainless steel. **18**(3), 538–545 (2015)
26. Mondal, D., Chattopadhyay, K., Singh, V.: Effect of ultrasonic shot peening on oxidation behavior of T91 and SS347 steels in air and steam at 650 °C. J. Mater. Eng. Perform. **29**(9), 5854–5870 (2020)

Chapter 21
Bainitic Transformation in Steel via Surface Mechanical Attrition Treatment at Room Temperature

Lokendra Kumar Katiyar, C. Sasikumar, and Mohd. Sarim Khan

Abstract Surface bainite transformation at room temperature is described when steel undergoes surface mechanically induced transformation. The steel surface is bombarded with high energy balls at ambient condition endures severe plastic deformation, mechanical alloying and mechanically induced phase transformation. The microstructural characterization showed the transformation of carbides interspersed within ferrite, i.e., signature of upper bainitic laths after 10 min. As the treatment time is increased from 10 to 40 min, the microstructure similar to a lower bainite that is formed because of twinning induced transformation that dominates at a higher strain rate. The first of a kind report on mechanically induced bainitic transformation through surface mechanical attrition treatment is witnessed with an order increase in dislocation density, i.e., from 2×10^{13} to 2×10^{14} in 40 min of treatment. Severe plastic deformation induced transformation occurred at the grain boundaries supplemented an increase in surface hardness to 275 VH (from 120 VH) and yield strength to 400 MPa (from 260 MPa).

Keywords Bainite · Stress induced transformation · Surface nanostructured steel · Severe plastic deformation

21.1 Introduction

Bainite is a non-equilibrium structure formed in steels when austenite is cooled continuously or isothermally just below the nose temperature of TTT diagrams. Charles E. Bain and E. S. Davenport first documented this microstructure in isothermal cooling of austenite [1]. Steels having a bainite microstructure that exhibits unique combination of mechanical properties such as higher strength without losing ductility, better hardness along with toughness and fatigue strength. The microstructure essentially consists of lath or plate like bainitic ferrite and carbide

L. K. Katiyar (✉) · C. Sasikumar · Mohd. S. Khan
Department of Materials and Metallurgical Engineering MANIT Bhopal, Bhopal, MP, India
e-mail: lokendrakatiyar90@gmail.com

A. Tewari et al. (eds.), *Proceedings of the 3rd International Conference on Advances in Materials Processing: Challenges and Opportunities*, Springer Proceedings in Physics 293, https://doi.org/10.1007/978-981-99-1971-0_21

Table 21.1 Chemical composition of HSLA steel used in the present investigation

C	Si	Mn	Cr	Ni	Cu	Al	Ti	V	Co	Fe
0.13	0.13	0.50	0.14	1.15	0.01	0.16	0.10	0.15	2.30	Bal.

precipitates as proposed by Bhadeshia et al. [2–4]. In the upper bainite, the carbide precipitates are occurring at grain boundaries and in lower bainite, the carbide precipitates are occurring within the bainitic ferrites. There are many reports on stress induced transformation to bainite reciting the role of stress in the transformation of austenite to bainite. The role of pre-stress or the stress applied during the transformation of austenite to bainite is investigated and the studies revealed a significant enhancement in the extent of bainitic transformations. The research work of Umemoto and co-workers showed that the transformation is accelerated if the applied stress exceeds the yield stress. The researchers have applied a tensile stress of 45–95 MPa during the transformation of austenite to bainite. Their results have demonstrated that the applied stress has a strong influence in the growth of bainite variants by contributing elongation along the longitudinal direction relative to the transverse direction. The present work deal with the role of severe plastic deformation stresses in transforming a ferrite into a bainite structure [5].

21.2 Experimental

21.2.1 Material

A high strength low alloy (HSLA) steel is used for the present investigation. The chemical composition of the steel is analyzed using a METAVISION atomic emission spectrometer and the results are shown in Table 21.1.

21.2.2 Surface Mechanical Attrition Treatment (SMAT)

The surface mechanical attrition treatment (SMAT) of steel samples was carried out using an industrial shot peening machine with the process parameters optimized for SMAT. The shot peening machine consists of a sample fixing stage, impeller to throw continuously the steel balls towards the samples, a suction unit to collect the balls/media after collision, dust cleaning chamber. The fixing table has a provision to rotate the samples so as to achieve uniform treatment. The working distance between table and the impeller is about 1 m and the throwing velocity is kept at 40 m/s.

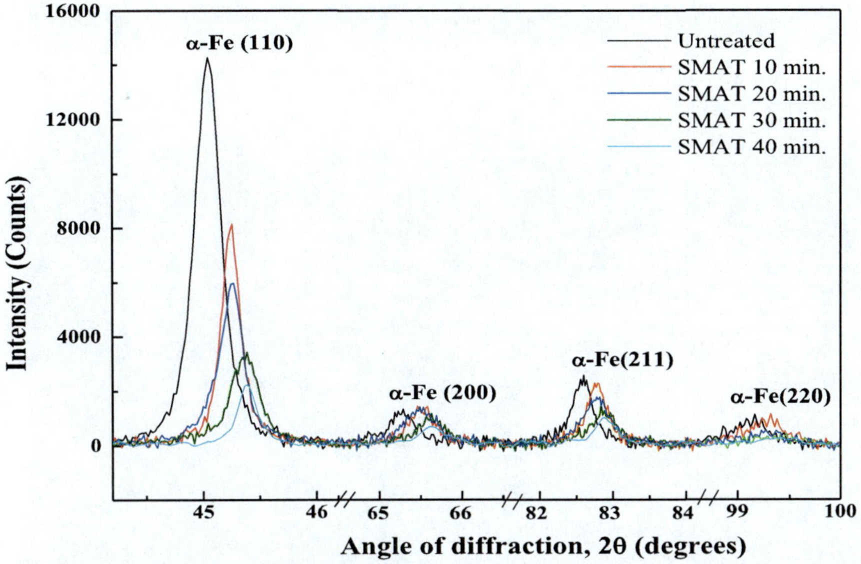

Fig. 21.1 X-ray diffraction pattern of samples before and after 10, 20, 30 and 40 min of SMAT respectively

21.3 Results and Discussions

21.3.1 XRD Analysis

As shown in Fig. 21.1, the xrd spectrum reveals characteristic peaks at about 45°, 65.2°, 82.6° and 99.2° corresponding to the reflections of (110), (200), (211) and (220) planes of ferrite (ICDD 00-006-0696). The peaks are shifted towards higher angle, broadened and their heights are reduced with time of surface mechanical attrition treatment.

21.3.2 Microstructure

The microstructure of the steel is obtained from the surface as well as the cross section to visualize the structural changes. Figure 21.2a–d shows the surface microstructures of the steel subjected to different time periods (0, 10, 20, 30 and 40 min) of the surface mechanical attrition treatments respectively. The surface microstructure of the steel prior to SMAT i.e., SMAT 0 shows equiaxed grains of polygonal ferrite and fine carbide precipitates (bright) within the ferrite grains and at the grain boundaries. The initial grain size is about 25 μm. The presence of micro-alloying elements such as

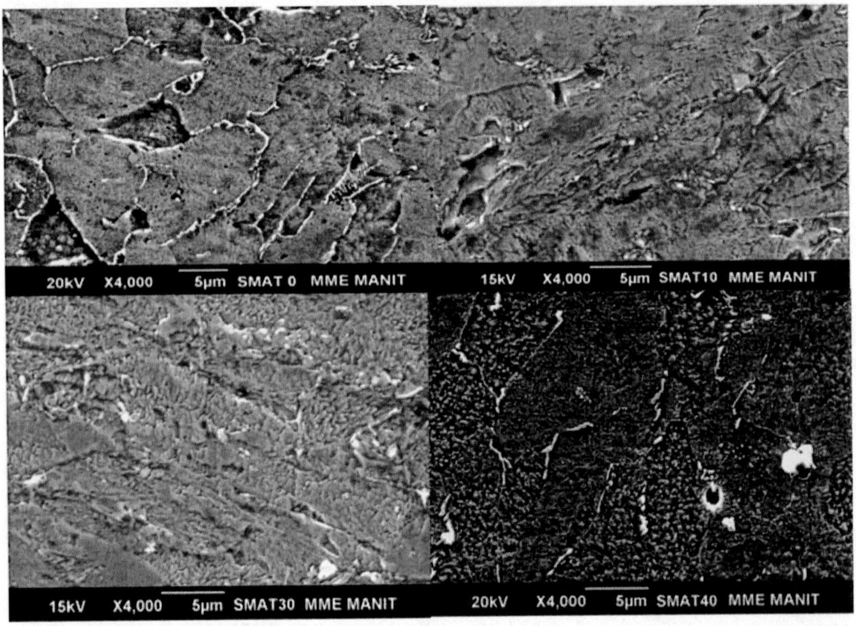

Fig. 21.2 Prior to SMAT, SEM images reveal equiaxed ferrite and carbide phases in the surface of the steel. **a** As received sample **b** SMAT for 10 min. **c** SMAT 30 min. **d** SMAT 40 min

Co, Mo, V, Nb, etc., not only decrease the interstitial carbon but also alter the pearlite structure into alloy carbides precipitated ferrite.

21.3.3 Hardness Profile

The surface hardness of the HSLA steel samples subjected to surface mechanical attrition for 0, 10, 20, 30 and 40 min are shown in Fig. 21.3. The surface hardness of the untreated steel is about 150 $HV_{0.05}$ and found increasing linearly with process time. A maximum of 275 $HV_{0.05}$ is obtained in the steel subjected to 40 min of SMAT. The hardness of thermally transformed bainite is found to be in the range of 250 to 300 HV as reported by other researchers. The microhardness obtained at selected ferrite phases revealed a hardness of 275 $HV_{0.05}$ while the carbide particles showed about 600 $HV_{0.05}$.

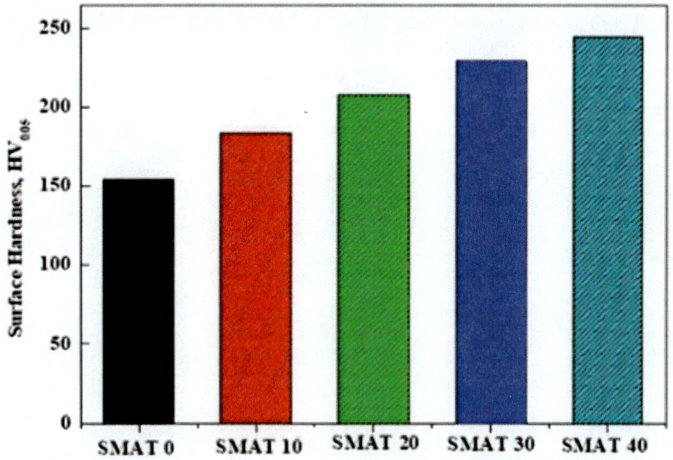

Fig. 21.3 The surface hardness of HSLA steel samples subjected to SMAT for different time periods

21.4 Conclusions

The surface mechanical attrition treatment (SMAT) causes severe plastic deformation of the steel surface, which then results in the transformation of equiaxed ferrite into lath and acicular ferrite and carbides precipitate along their boundaries, indicating an upper bainite phase. In the course of treatment, the ferrite grains are transformed into acicular ferrite and carbide precipitates within them, resulting in a lower bainite.

The transformation of equiaxed ferrite into bainitic ferrite resulted in shifting of x-ray peaks and the nano-crystallization has caused broadening. The crystallite size decreased to about 50 nm and the average dislocation density of the crystals increased exponentially from 2×10^{13} to 2×10^{14} per square meter.

The hardness of the treated samples was found varying across the cross section. A maximum surface hardness of 275 $HV_{0.1}$ is obtained at the surface while the core showed a hardness of 150 $HV_{0.1}$.

References

1. Davenport, E.S., Bain, E.C.: Transformation of austenite at constant subcritical temperatures, Metall Trans. 1(1970) 3503–3530. https://doi.org/10.1007/BF03037892.
2. Bhadeshia, H.K.D.H., Christian, J.W.: Bainite in steels. Metall. Trans. A. **21**, 767–797 (1990). https://doi.org/10.1007/BF02656561.

3. Fielding, L.C.D.: The bainite controversy. Mater. Sci. Technol. **29**, 383–399 (2013). https://doi.org/10.1179/1743284712Y.0000000157.
4. Bhadeshia, H.K.D.H.: Bainite in Steels: Transformations, Microstructure and Properties, 2nd edn. IOM Communications, London (2001)
5. Bhadeshia, H.K.D.H.: Interpretation of the microstructure of steels. At. Mech. Transform., 1–37 (2008)

Chapter 22
Numerical Investigation of Weld Induced Residual Stress State Field for Improved Structural Integrity in AUSC Power Plant Applications

Pavan Meena and Ramkishor Anant

Abstract Recently, computational simulation for residual stresses formation in martensitic and austenitic stainless steel has received great attention in the industry due to its wide range of applications in petroleum refinement stations, superheater and reheater tubing in nuclear, thermal and other advanced ultra-supercritical (AUSC) power plants. In order to integrate martensitic steel components and austenitic stainless steel piping systems, dissimilar welded connections are frequently used in fossil power stations. Such welded structures' structural integrity is dependent on residual stresses induced during joining process. Numerous studies have been conducted to examine residual stress phenomena and its effect on the mechanical characteristics of the components. Several techniques have been developed over the time to assess residual stress for various kinds of components in order to provide a reliable assessment. This computational study aims to compute thermal histories at different locations in heat affected zones (HAZ) of weldment, prediction of weld induced residual stresses, and to provide an overview of some of the recent advances in this approach to help researchers on selecting their techniques.

Keywords Residual stresses · FEA · AISI304 & AISI316L

22.1 Introduction

The structural integrity of welded structures is mainly depending on various process parameters, thermo-mechanical properties and microstructure of the materials to be joined. However, residual stresses induced by the joining process lead to stress corrosion cracking in case of stainless steels and affect it joint integrity [1]. During welding, synchronous irregular heating and cooling, and shrinkage owing to varied cooling rates at various areas of the weld joint, and stresses created by metallurgical

P. Meena (✉) · R. Anant
Department of Materials and Metallurgical Engineering, Maulana Azad National Institute of Technology, Bhopal, MP 462003, India
e-mail: 203119005@stu.manit.ac.in

A. Tewari et al. (eds.), *Proceedings of the 3rd International Conference on Advances in Materials Processing: Challenges and Opportunities*, Springer Proceedings in Physics 293, https://doi.org/10.1007/978-981-99-1971-0_22

phase change are the main causes of residual stress in dissimilar weld joint. The solid-state phase change and solidification cause the stresses to be produced. The mismatch between various sections and phases also contributes to the generation of residual stresses. Tensile residual stress (typically present at the weld pool) and compressive residual stresses (away from the weld line) are produced during the welding process [2].

The welded joints in power plants may prematurely fail due to residual stresses that are produced during the welding process and these stresses depend on various factors as shown in Fig. 22.1c. Therefore, in-depth research is needed on the residual stresses induced in dissimilar weld joints. The performance of the welded structures is affected by residual stresses in an assortment of ways, including brittle fracture, fatigue failure, and stress corrosion cracking. For welded structures to be used safely, a detailed assessment of residual stresses is essential. One of the best techniques for figuring out residual stresses is the finite element approach. The weld induced residual stresses are also determined using a wide range of experimental techniques. These unwanted residual stresses might lead to catastrophic fracture of the component while in operation. The components fatigue life is decreased by the tensile residual stresses that also encourage brittle fracture, buckling, and stress corrosion cracking. Different welding process causes residual stresses in fusion and other heat affected zones. Typically, the residual stress value at this region is greater than the materials yield stress, leading to component failure [3]. Experimental techniques such as centre hole drilling and deep hole drilling are categorized as destructive techniques. The finite element method (FEM) using in the ANSYS, WELDSIM, SIMUFACT, VFT, SYSWELD and ABAQUS software comes in non-destructive techniques. Neutron, ultrasonic, and XRD techniques are the other physical experimental based non-destructive approach [1].

Considering the FEM approach with other techniques like deep hole drilling and the diffraction method, which can only identify residual stresses at a specific location, the advantage of using the FEM approach is that it is able determine residual stresses across the entire joint structure. The finite element technique (FEM) is more frequently employed since it can calculate residual stress across the entire structure. This approach is also the most accurate, but it has the drawback that the outcomes of the process depend on the veracity of the data used as input. Figure 22.1a shows schematic illustration of computational procedure for residual stress induced in weldment, weld joint configuration and effect of various parameters on residual stress formation.

22.2 Computational Set-Up and Measurements

The weld sample consisted of two 8 mm thick plates of austenitic stainless-steel series (AISI304 & AISI316L), each of dimensions of 150 mm × 65 mm (Fig. 22.1b), that are welded into a v-groove butt joint with root gap and face of 1.5 mm in single passes within AWI by using ABAQUS/CAE standard software [4].

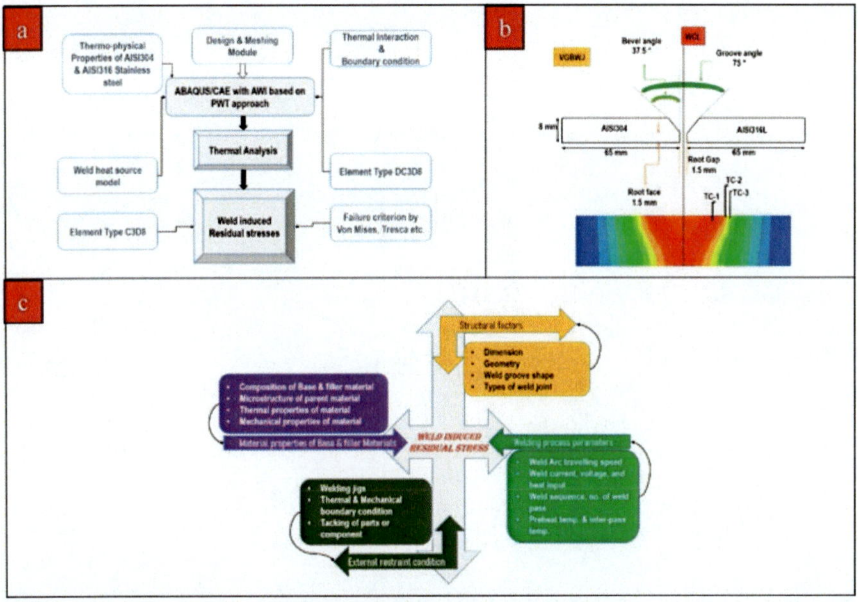

Fig. 22.1 Schematic illustration **a** computational procedure for welding simulation, **b** weld joint configuration, and **c** effect of various parameters on residual stress formation

The governing finite element equation used in this computational model is

$$CdT/dt + kT - f_{\mathrm{T}} \tag{22.1}$$

here c, k are temperature dependent heat capacity and conductivity matrix, f_{T} is thermal load vector.

22.2.1 Prescribed Temperature via Gaussian Double Ellipsoidal Model

The double ellipsoidal model is represented by the following equation [4]

$$\frac{x^2}{a^2} + \frac{y^2}{b^2} + \frac{z^2}{c^2} \le 1 \tag{22.2}$$

Correspondingly, if the nodes are prescribed with temperature [5]

$$T = T_0 \exp\left\{ B\left(\frac{x^2}{a^2} + \frac{y^2}{b^2} + \frac{z^2}{c^2}\right)\right\} \tag{22.3}$$

$$\text{If} \quad \frac{x^2}{a^2} + \frac{y^2}{b^2} + \frac{z^2}{c^2} = 1 \tag{22.4}$$

$$T_\text{m} = T_0 e^B \tag{22.5}$$

$$B = \ln\left[\frac{T_\text{m}}{T_0}\right] \tag{22.6}$$

Substitute Eq. (22.6) into Eq. (22.3) [5]

$$T = T_0\left(\frac{T_\text{m}}{T_0}\right) \tag{22.7}$$

The generalised heat conduction in solid body is given in Eq. (22.6) [6].

$$\frac{\partial}{\partial x}\left(k_x \frac{\partial T}{\partial x}\right) + \frac{\partial}{\partial y}\left(k_y \frac{\partial T}{\partial x}\right) + \frac{\partial}{\partial z}\left(k_z \frac{\partial T}{\partial x}\right) + Q = \rho C \frac{\partial T}{\partial t} \tag{22.8}$$

22.3 Results and Discussion

In principle, metal is heated to a melting point during the welding process, after which it is allowed to cool and solidify, forming beads in the area where the two steel plate halves are joined. Herein, coupled thermo-mechanical procedure has adopted that involves thermal analysis (transient temperature and phase evolution) and mechanical analysis (used temperature history of all nodes from FE model). The transient temperature field sturdily affects the residual stress state field. Moreover, the computational model has already verified with the previous published work in the similar context for both thermal and mechanical analysis.

Figure 22.2b and c illustrate tensile and compressive residual stress along with all principal stresses over surface of the dissimilar welded joint. The transverse residual stresses over the surface of weldment have similar trends of profile observed at both the side of weld centre line due to self-balancing nature of residual stresses. All the stresses are computed along the PATH selected at mid of the weldment in transverse direction. However, thermal cycle curve obtained from the simulation at different locations in heat affected zones as shown in Fig. 22.1b.

Figure 22.3a shows thermal cycle curve where temperature is a function of time at some specified nodes in different locations of heat affected zone (HAZ). Moreover, weld induced residual stress distribution over the surface of weldment of AISI304/316L in transverse direction along the path selected by considering all nodes at mid of weldment is shown in Figs. 22.2b, c and 22.3b.

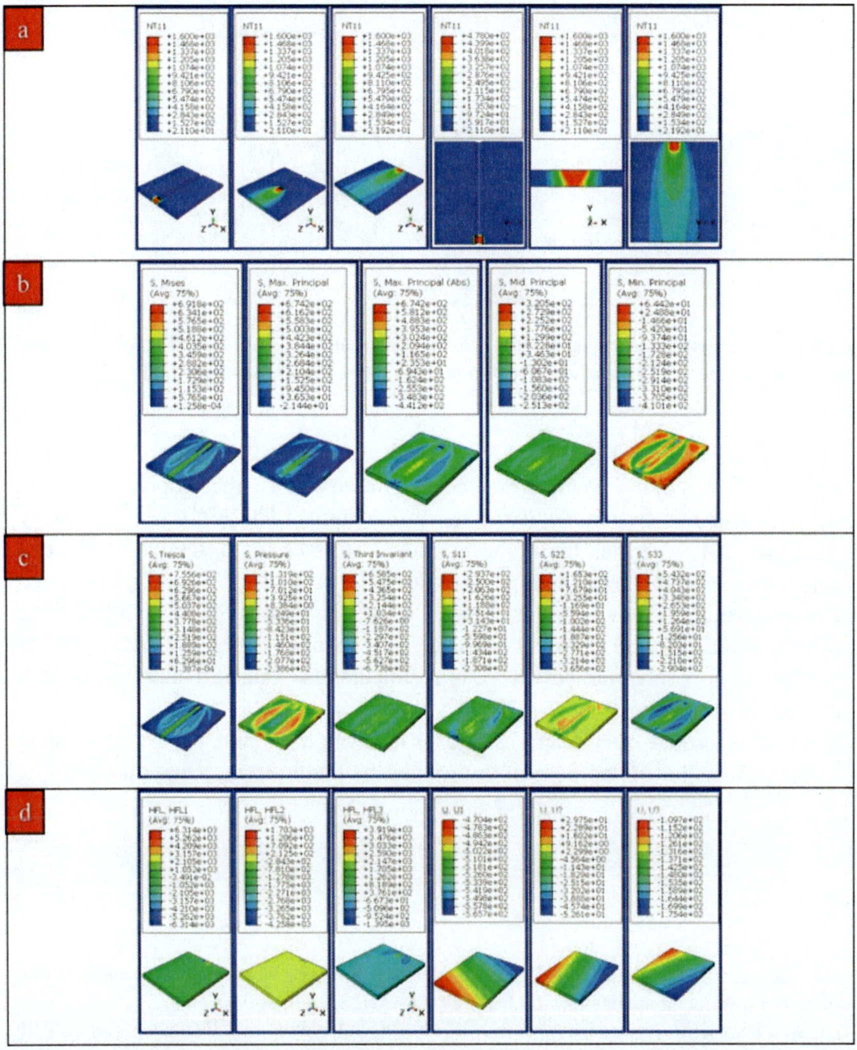

Fig. 22.2 Depicted **a** Weld torch movement and Nodal temperature variation **b** von Mises and other principal stresses, **c** Tresca, pressure, Third invariant, S11, S22, S33 stresses, **d** heat flux and Deflection distribution field in *x*, *y*, and *z*-direction

22.4 Conclusions

This computational study uses element birth and death technique, which is used to model the welding process, to compute weld induced residual stress in AISI304/AISI316L dissimilar welded joints. The following conclusions can be drawn from the results:

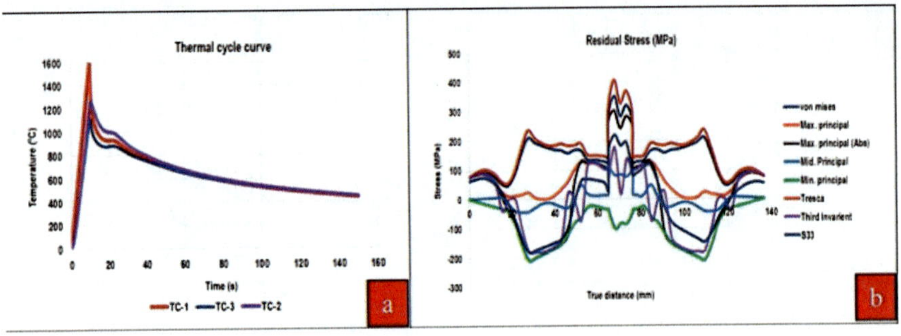

Fig. 22.3 **a** Thermal cycle curve at specified locations, and **b** weld induced residual stresses distribution along the PATH selected at mid of the weldment in transverse direction.

- The residual stress distribution in 8 mm thick AISI304/AISI316L dissimilar plate weld joint is majorly symmetric due to similarities in thermophysical material properties of two sides of the dissimilar joints. The HAZ region of both AISI304/316L steel has the highest residual stresses (compressive or tensile), which may cause mismatched welds to fail in this region.
- The groove angle has significant effect on residual stress. The residual stress distribution across the erratically welded joint changes extremely little when the groove angle is lowered, yet the peak residual stress values remain the same at the specified spots.
- Standard v-groove specifications and groove angle should be used to reduce residual stress for the purpose of improving the joints strength and durability.

References

1. Dak, G., Pandey, C.: A critical review on dissimilar welds joint between martensitic and austenitic steel for power plant application. J. Manuf. Process. **1**(58), 377–406 (2020)
2. De, A., Debroy, T.: A perspective on residual stresses in welding. Sci. Technol. Weld. Join. **16**, 204–208 (2017)
3. Yaghi, A.H., Hyde, T.H., Becker, A.A., Sun, W.: Finite element simulation of residual stresses induced by the dissimilar welding of a P92 steel pipe with weld metal IN625. Int. J. Press. Vessel. Pip. **111–112**, 173–86 (2013). ijpvp.2013.07.002
4. Seleš, K., Perić, M., Tonković, Z.: Numerical simulation of a welding process using a prescribed temperature approach. J. Constr. Steel Res. **1**(145), 49–57 (2018)
5. Goldak, J., Akhlaghi, M.: Computational Welding Mechanics. Springer, New York Inc., USA, ISBN 9780387232874 (2005)
6. Meena, P., Kumar, M., Singh, M., Kumar Shukla, D., Burela, R.G., Jhunjhunwala, P., Gupta, A., Pandey, C.: Numerical analysis of different SUS304 steel weld joint configurations using new prescribed temperature approach. Trans. Indian Inst. Met. **75**(6), 1649–1668 (2022)

Chapter 23
Experimental and Numerical Analyses of Uniaxial Tensile Test of Automotive Grade Anisotropic Sheets

Amit Kumar and Shamik Basak

Abstract Formability is an important aspect of sheet metal forming industries. Numerical simulation is often used to predict the formability of the sheet metal. In the present study, experimental and numerical analyses of the uniaxial tensile test were conducted to predict the mechanical properties of extra deep drawing (EDD) steel sheets. In experimental analyses, tensile test specimens were machined along the three orientations, with the tensile axis being parallel, diagonal and perpendicular to the rolling direction of the sheet, to assess the value of anisotropy constants along three directions. In the numerical approach, LS-Dyna software was used to simulate the experiment. Two different element formulations, such as solid and shell elements, were used for discretizing the input blanks. Further, two different yield criteria, namely isotropic von Mises and anisotropic Hill48, were used as the material models. The results of both approaches were compared in terms of load–displacement curve, yield point, ultimate tensile strength point, and failure point. The analyses showed that there is a good agreement between the experimental and simulation results.

Keywords Plastic anisotropy · Uniaxial tensile tests · FE simulation

23.1 Introduction

Extra deep drawing (EDD) steel has gained large attention in automotive industries due to its constant use in fabricating different automotive parts, owing to its excellent formability. To predict formability, finite element (FE) simulations play a crucial role by reducing the cost of manufacturing and the lead time. The predictive accuracy of the formability in FE simulations is primarily dependent upon the

A. Kumar · S. Basak (✉)
Department of Mechanical and Industrial Engineering, Indian Institute of Technology Roorkee, Roorkee 247667, India
e-mail: shamik.basak@me.iitr.ac.in

A. Tewari et al. (eds.), *Proceedings of the 3rd International Conference on Advances in Materials Processing: Challenges and Opportunities*, Springer Proceedings in Physics 293, https://doi.org/10.1007/978-981-99-1971-0_23

quality of the input material properties in the simulation. Hence, experimental evaluation and subsequent theoretical prediction of the material properties, along with work hardening behavior, is of prime interest to the researchers [1]. Further, the anisotropy developed during the cold rolling of the sheet metal plays a vital role in the prediction of formability [2]. Consequently, various anisotropic yield models were proposed to predict the material properties using the anisotropic coefficients [3, 4]. However, there is a dearth of rigorous numerical studies where the FE data is used to predict the material response under basic formability tests such as the uniaxial tensile test. Therefore, in this study, uniaxial tensile tests of the EDD sheet materials were conducted along various anisotropic orientations. Accordingly, the influence of anisotropy in the experimental data prediction was studied using the isotropic von Mises and anisotropic Hill48 model [5, 6] in the FE simulation. Furthermore, the solid and shell element formulation strategies were used to identify the failure location and effective plastic strain contours from the FE simulation.

23.2 Experimental Procedure

In the present work, 1 mm thick EDD steel sheet was prepared for the uniaxial tensile test as per ASTM E8M standard using wire-cut EDM process. The specimens were cut along three directions viz. 0°, 45°, and 90° with reference to the rolling direction (RD) of the sheet to measure the anisotropy of the material. To measure the Lankford anisotropy coefficients (r-values) along three different directions, the specimens were elongated up to 70% of the ultimate tensile strength (UTS), and the corresponding strain level was calculated using circular grid analysis method [7]. The r-values were calculated using Eq. (23.1).

$$r = \varepsilon_w/\varepsilon_t = -\varepsilon_w/(\varepsilon_l + \varepsilon_w) \tag{23.1}$$

here, ϵ_w, ϵ_l and ϵ_t represent the true strains along the width, length, and thickness directions, respectively. Experiments were conducted using the universal testing machine with an initial strain rate of 0.001 s^{-1}. The mechanical properties evaluated during the test are reported in Table 23.1 below.

Table 23.1 Mechanical properties of the EDD sheet metal

Orientations	Yield strength (MPa)	Ultimate tensile strength (MPa)	Lankford's coefficients (r-values)
RD	232.1 ± 3.9	328.2 ± 2.7	1.54 ± 0.05
45° to RD	222.3 ± 5.1	317.3 ± 6.1	1.66 ± 0.05
90° to RD	251.4 ± 4.1	338.6 ± 7.8	1.74 ± 0.07

23.3 Numerical Procedure

The success of numerical simulation lies in how well the simulation results match with the experimental results. This is only possible if the accurate material properties along with the proper material models are selected. In the present work, FE-based commercial software named LS-Dyna was used to simulate the uniaxial tensile test. The foremost step for the simulation study is the model formulation. Therefore, two different element formulations, such as solid and shell elements, were used for discretizing the input blank. The nodes on one end of the blank were fixed, while the nodes on the other end were assigned a predefined velocity profile. Further, to understand the effect of the anisotropy of the sheet material on the formability prediction, two different isotropic and anisotropic-based yield models were used. Subsequently, different FE simulations were conducted by incorporating the material model as isotropic von Mises yield criterion and anisotropic Hill48 yield criterion. The anisotropy coefficients were used as an input parameter to Hill's material model. The anisotropic Hill48 yield function in plane stress condition can be represented as per Eq. (23.2).

$$\phi(\sigma) = \overline{\sigma}^2 = F\sigma_{YY}^2 + G\sigma_{XX}^2 + H(\sigma_{XX} - \sigma_{YY})^2 + 2N\sigma_{XY}^2 \quad (23.2)$$

Here, F, G, H, and N described the anisotropy of the sheet metal where $F = \frac{r_0}{r_{90}(1+r_0)}$, $G = \frac{1}{1+r_0}$, $H = \frac{r_0}{1+r_0}$, and $N = \frac{(r_0+r_{90})(1+2r_{45})}{2r_{90}(1+r_0)}$. Numerical values of the anisotropic constants (r-values) were calculated and shown in Table 23.1. For isotropic sheet materials, the anisotropic constants of the Hill48 model are reduced to $F = G = H = 0.5$, and $N = 1.5$.

23.4 Results and Discussion

The mechanical properties of the EDD sheets along various anisotropic directions are shown in Table 23.1. Significant differences are observed in yield strength, UTS, and r-values along different orientations. This is due to the change in the crystallographic texture during the cold rolling process, for which the sheet metal exhibits significant anisotropy in the mechanical properties. Therefore, the objective was to analyze the effect of anisotropy on the load progression response during the uniaxial tensile test. Accordingly, four different FE models of uniaxial tensile test were simulated. Among them, two material models were isotropic von Mises models, and another two were anisotropic Hill48 models. For each of the isotropic/anisotropic material the models were further assigned with shell and solid element formulations. The predicted load progression response along RD of the sheet metal was recorded and plotted with the experimental data in Fig. 23.1. It can be observed that the anisotropic Hill48 model (both shell and solid) predicts the experimental load progression curve better than the isotropic model. Moreover, solid element formulation with the Hill48 model exhibits

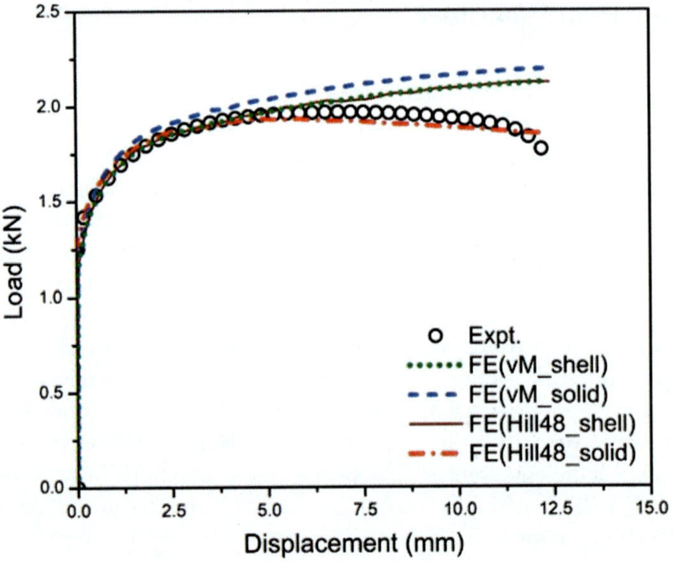

Fig. 23.1 Comparison of the experimental load–displacement curve with simulation results incorporating various yield criteria with shell and solid element models

the best prediction of the load progression curve for the uniaxial tensile test. Hence, it can be concluded that the anisotropic model is necessary for the precise prediction of the formability of the sheet metal.

Further, the deformed FE sample at the onset of failure is shown in Fig. 23.2. The experimental load–displacement curve along RD is shown in Fig. 23.2a, indicating point A, which corresponds to the specimen at an elongation of 12 mm experimentally. The accurate deformation step was identified from the best predicted FE model (i.e., Hill48 material model with solid element formulation), and the deformed FE specimen is shown in Fig. 23.2b. Clearly, the localized neck for the deformed FE sample is visible in the figure, which indicates that the sample is at the onset of the failure point. To validate the FE model, the elongated gauge length was calculated, as shown in Fig. 23.2b, and it can be observed that the elongated gauge length is 36.7 mm (24.8 + 11.9). Therefore, the total elongation is 11.7 mm from the FE simulation, which corroborates well with the experimental data of 12 mm elongation in Fig. 23.2a. This indicates an efficient FE model for the uniaxial tensile test.

The deformation behavior of the uniaxial tensile specimen was studied corresponding to the Hill48 material model with solid element and shell element formulations in terms of effective plastic strain contour, as shown in Fig. 23.3. Although the effective plastic strain values for both the solid and shell elements are comparable, still a major difference in necking location can be observed in the FE simulation. It can be observed from Fig. 23.3 that the failure location i.e. necking, takes place nearly at the center of the gauge section in the case of solid element, whereas the necking zone shifts toward the grip section in the case of shell element formulation.

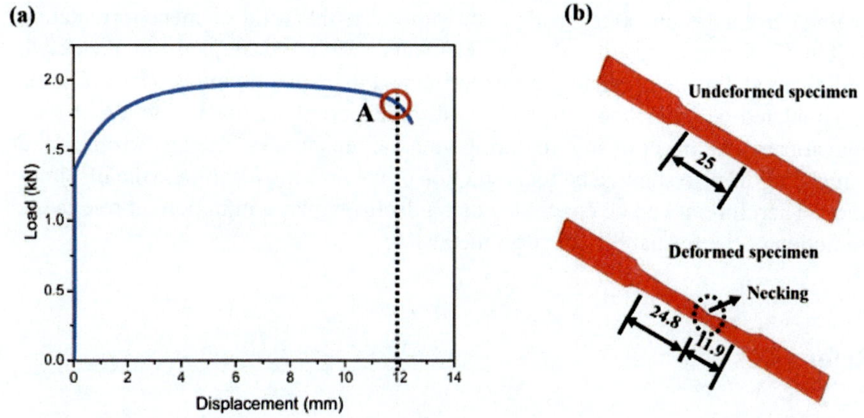

Fig. 23.2 **a** The experimental load–displacement curve along with **b** FE simulated samples at the undeformed condition and deformed condition up to the displacement indicated as A

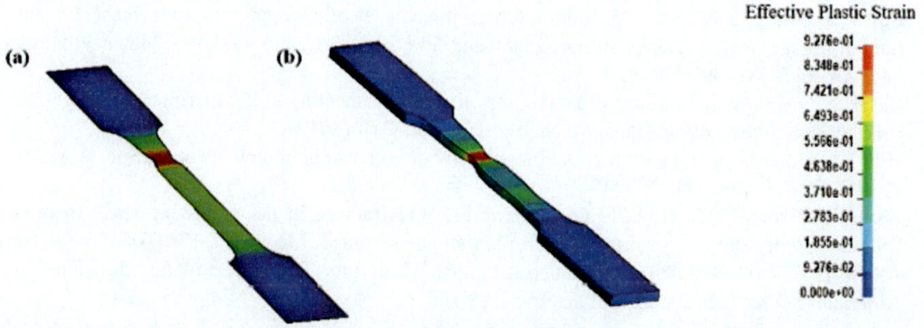

Fig. 23.3 Effective plastic strain contour of tensile test specimen with **a** shell element and **b** solid element meshing strategy using Hill48 anisotropic material model

Hence, it can be concluded from the above discussions that the solid element formulation gives better results with respect to the shell element formation as far as the location of the localized necking appears.

23.5 Conclusions

In this work, uniaxial tensile tests of EDD steel sheets along various anisotropic directions were conducted. Accordingly, the experimental load progression data along the rolling direction was predicted from FE simulations implementing both isotropic and anisotropic material models with solid and shell element formulations. It was observed that the anisotropic Hill48 material model, in comparison to the isotropic von Mises material model, exhibits better prediction of the load progression data obtained from the uniaxial tensile test. Moreover, the solid element formulation exhibited better accuracy than the shell element formulation coupled with the Hill48 material model while predicting the experimental load–displacement data. Also, the FE simulated data corroborated well with the experimental data in terms of total elongation at the onset of failure during uniaxial tensile test. The anisotropic solid element model also shows a better prediction of the necking location in the FE simulation. Therefore, it can be concluded that anisotropy plays an important role in the prediction of the formability of sheet metals.

References

1. Holmberg, S., Enquist, B., Thilderkvist, P.: Evaluation of sheet metal formability by tensile tests. J. Mater. Process. Technol. **145**(1), 72–83 (2004)
2. Basak, S., Kim, C., Jeong, W., Jung, Y.I., Lee, M.G.: Numerical prediction of sheared edge profiles in sheet metal trimming using ductile fracture modeling. Int. J. Mech. Sci. **219**, 107109 (2022)
3. Basak, S., Panda, S.K.: Necking and fracture limit analyses of different pre-strained sheet materials in polar effective plastic strain locus using Yld 2000–2d yield model. J. Mater. Process. Technol. **267**, 289–307 (2019)
4. Basak, S., Panda, S.K.: Failure strains of anisotropic thin sheet metals: experimental evaluation and theoretical prediction. Int. J. Mech. Sci. **151**, 356–374 (2019)
5. Hill, R.: A theory of the yielding and plastic flow of anisotropic metals. Proc. R Soc. A Math. Phys. Eng. Sci. **193**, 281–297 (1948)
6. Basak, S., Panda, S.K., Lee, M.G.: Formability and fracture in deep drawing sheet metals: extended studies for pre-strained anisotropic thin sheets. Int. J. Mech. Sci. **170**, 105346 (2020)
7. ASTM E. 517-00. Standard test method for plastic strain ratio R for sheet metal. Philadelphia: American Society for testing and materials. (2006).

Part III
Materials for Sustainability

Chapter 24
Green Graphene-Based Anticorrosive Coating for Carbon Steel in Marine Environment

Anu Verma, Chandra Sekhar Tiwary, and Jayanta Bhattacharya

Abstract The use of green graphene (GG), as a green corrosion-resistant material for carbon steel, has recently gained much attention in the scientific community. We have prepared a green graphene/epoxy (GG/EP) based anticorrosive coating for the carbon steel. However, green graphene (GG) was synthesized from carbonaceous agricultural biomass. The corrosion inhibition property of GG/EP coating on carbon steel in 3.5 wt% NaCl solution was investigated using the potentiodynamic polarization technique. The results showed a decrease in corrosion rate up to 92.9% compared to uncoated carbon steel coupons. The SEM images of GG/EP coating show uniform distribution and finish surface on the carbon steel substrate.

Keywords Green graphene · Carbon steel · Anticorrosion coating · Electrochemical techniques

24.1 Introduction

Corrosion is a major global problem for metals industries that causes significant losses. The National Association of Corrosion Engineers (NACE International), a global organization, conducted research on the impact of US government policy and industry on corrosion management. Corrosion is predicted to have a global cost of \$2.5 trillion, equal to 3.4% of the US GDP and around 4% in India. Corrosion-related

A. Verma (✉) · J. Bhattacharya
School of Environmental Science and Engineering, Indian Institute of Technology, Kharagpur, West Bengal 721302, India
e-mail: anuverma0168@iitkgp.ac.in

J. Bhattacharya
Department of Metallurgical and Materials Engineering, Indian Institute of Technology, Kharagpur, West Bengal 721302, India

C. S. Tiwary
Department of Mining Engineering, Indian Institute of Technology, Kharagpur, West Bengal 721302, India

A. Tewari et al. (eds.), *Proceedings of the 3rd International Conference on Advances in Materials Processing: Challenges and Opportunities*, Springer Proceedings in Physics 293, https://doi.org/10.1007/978-981-99-1971-0_24

disruptions, such as safety concerns and plant outages, have a detrimental effect on human life and the economy. To increase the lifespan and dependability of metal components currently in use, corrosion inhibition for metal structures is a crucial step [1]. Chromate-based coatings, including primers and pigments, have proven to be the most effective method for preventing corrosion over time [2]. However, due to environmental problems and people's health brought on by the emissions of volatile organic compounds, they must be replaced with more environmentally friendly technologies. In the previous few decades, extensive research has been done to create new, smart, and environmentally friendly corrosion inhibitor coatings, which has led to a major hunt for water-based alternatives [3–5]. Because they are biodegradable and free of heavy metals and other hazardous substances, green inhibitors are one of the finest solutions for preventing corrosion in metals and alloys. The smart coating category is a well-known inhibitory product in the coatings industry. These intelligent coatings, which include inhibitors and other additives, reflect advancements in corrosion prevention as well as their multifunctionality and advantageous effects on the environment. They have auto-responsive traits that cause them to react quickly when they are exposed to harsh environments.

In this research, we have prepared a green graphene/epoxy (GG/EP) based anti-corrosive coating for carbon steel. However, green graphene (GG) was synthesized from carbonaceous agricultural biomass. The corrosion inhibition property of GG/EP coating on carbon steel in 3.5 wt% NaCl solution was investigated using the potentiodynamic polarization technique. The results showed a decrease in corrosion rate up to 92.9% compared to uncoated carbon steel specimens. SEM micrographs of the GG/EP-coated carbon steel surfaces show uniform. These coatings are more resistant to oxidation and corrosion in various corrosive environments than uncoated metals.

24.2 Experimental Section

24.2.1 Materials

Chemicals such as Hydrochloric acid (37%), Acetone (99.9%), Distilled water, Ethanol (99.9%), and Isopropanol (60%) were of regent's grade and purchased from merk company. Sodium chloride (99.5%), Potassium Chloride (98.5%), and Lab grade Epoxy resin and Hardener were purchased from Sigma-Aldrich, USA. Green graphene (GG) was synthesized from a mixture of carbonaceous agricultural biomass on a lab scale.

For the electrochemical test, carbon steel panels were employed with the elemental composition of C (0.09 wt%), Mn (0.56 wt%), Si (0.18 wt%), S (0.12 wt%), P (0.10 wt%), Cr (0.01 wt%), Ni (0.03 wt%), Cu (0.05 wt%), Ti (0.01 wt%), and Fe (98.83 wt%).

24.2.2 Preparation of Emulsion and Working Electrode

Prepared green graphene (GG) was mechanically dispersed in a mixture of solvents (Acetone, Ethanol, DMF, Isopropanol, Xylene, etc.). It was then combined with a polymeric matrix (epoxy resin) using starrer mechanically beginning at 100 rpm followed by 200, 400, and 600 rpm for 20 min at room temperature. The prepared emulsion was left for a few minutes to remove any air bubbles. The final step was to combine the produced emulsion with a curing agent that cures solid coatings by adding 1 g per 3 g of epoxy resin at room temperature. Finally, the emulsion was applied to carbon steel panels using the dip-coating method.

The panels were cut into desired dimensions of 10 mm, 10 mm, and 2 mm in length, width, and thickness, respectively. One surface of the coated and uncoated fabricated sample was covered with inert tape so that the exposed area was fixed to 1 cm^2. These samples were then taken for electrochemical experiments.

24.2.3 Electrochemical Measurement

The anti-corrosion performance was conducted on the uncoated and GG/EP-coated carbon steel electrode using electrochemical techniques, i.e., Potentiodynamic polarization (PDP), compared to the uncoated carbon steel electrode in a 3.5 wt% NaCl solution. Three electrode standard corrosion cell was used for corrosion measurements. Platinum was used as a counter electrode, Ag/AgCl as a reference, and carbon steel acted as working electrodes. Potentiodynamic polarization was performed with a potential range of −250 mV to +250 mV and a scan rate of 1 mV/s. The electrochemical cell was filled with a 3.5 wt% NaCl solution followed by open circuit potential (E_{ocp} measurement). The resulting curves of the potentiodynamic polarization were analyzed using the software (Z' view), from which all the corrosion parameters were predicted and listed (see Table 24.1). The corrosion inhibition efficiency was calculated using Eq. 24.1, where I^0 and I are the corrosion rates of uncoated and coated carbon steel electrode specimens, respectively.

$$\text{IE}\ (\%) = \frac{(I^0 - I)}{I^0} \times 100 \quad (24.1)$$

Table 24.1 E_{corr}, I_{corr}, corrosion rate, and corrosion inhibition efficiency obtained from polarization curves in 3.5 wt% NaCl solution

Sample	E_{corr} (V)	I_{corr} (A/cm^2)	Corrosion rate (mmpy)	Inhibition efficiency (%)
Uncoated (CS)	−0.360	4.001	0.309	–
Epoxy (EP)	−0.511	2.437	0.188	39.1
GG/EP	−0.228	2.819	0.021	92.9

24.3 Results and Discussion

24.3.1 Potentiodynamic Polarization Study

A polarization plot (see Fig. 24.1) was used to estimate the partial electrode reactions, and the corrosion potential (E_{corr}) and accompanying corrosion current density were calculated (I_{corr}). Tafel cathodic and anodic constants, denoted by the letters c and a, were obtained using linear fitting of the Tafel slopes. All of the metals' corrosion potentials shifted to significantly higher positive values, which could indicate an anodic kind of inhibitor [6]. GG/EP can be considered a mixed-type inhibitor because it prevents corrosion by geometrically blocking the carbon steel's cathodic and anodic surface-active sites.

The polarization curves for the uncoated and coated with EP and GG/EP on carbon steel in 3.5 wt% NaCl solutions are shown in (see Fig. 24.1), respectively. The electrochemical parameters given in (see Table 24.1) are those predicted by the polarization studies shown in (see Fig. 24.1). The polarization results indicate that as the I_{corr} value decreases, the corrosion rate of the GG/EP-coated carbon steel decreases compared to uncoated carbon steel specimen. The carbon steel specimen coated with GG/EP shows the highest corrosion inhibition efficiency of 92.9% and shows more homogeneous corrosion resistance than the uncoated sample.

The uncoated carbon steel specimen exhibits a stable oxide film over a wide range of potentials in the cathodic region during cyclic polarisation in a 3.5 wt% NaCl solution. Still, it also shows extremely high corrosion potential E_{corr} (−0.360 V) and corrosion current I_{corr} (4.001 A/cm^2) compared to the other coated specimens. The specimen coated with EP and GG/EP had the lowest corrosion current density (2.437 A/cm^2 and 2.819 A/cm^2) compared to the uncoated carbon steel specimen (see Fig. 24.1). The inhibition efficiency of the GG/EP-coated carbon steel specimen was found to be 92.9% higher as compared to the uncoated carbon steel specimen. Thus, GG/EP coated specimen demonstrated enhanced corrosion resistance even at

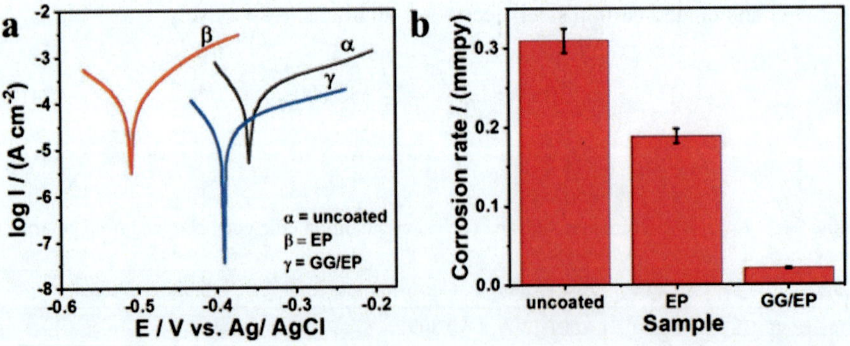

Fig. 24.1 **a** Polarization plot for uncoated, EP coated, and GG/EP coated carbon steel specimen **b** corrosion rate versus sample

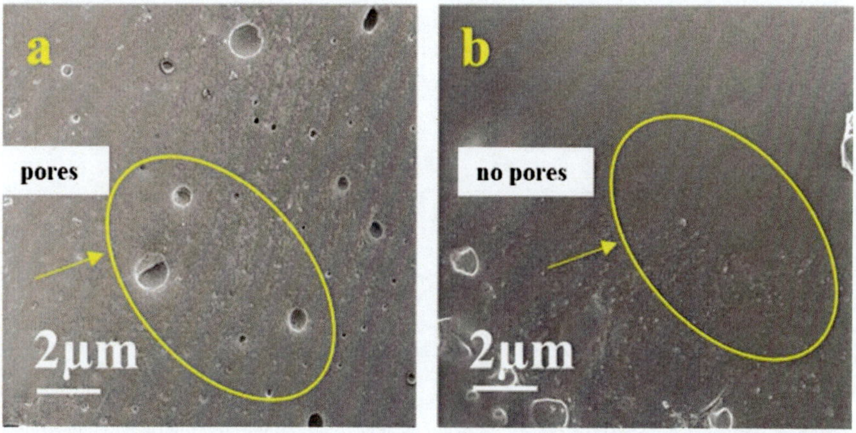

Fig. 24.2 Scanning electron microscopy images of **a** EP coated, and **b** GG/EP coated carbon steel specimen

a low concentration of 1 wt% in a 3.5 wt% NaCl solution. From the polarization plot, it can be attributed that the reduction in corrosion rate is mainly due to the increase in reduction reaction and decrease in the anodic current density. This implies that the coating will prevent the anodic dissolution of the carbon steel in the electrolyte. The increasing trend of E_{corr} also confirms this observation.

24.3.2 Surface Morphology and Protection

Scanning electron micrographs of the EP and GG/EP coated carbon steel surfaces are shown in (see Fig. 23.2a and b). The EP-coated carbon steel surface shows hydrophilic behavior with large pores. However, GG/EP coated carbon steel surface shows uniform coating without pores. Examining the photos in the right-hand column demonstrates that the GG offered effective protection for the carbon steel under consideration.

24.4 Conclusions

Compared to uncoated, GG/EP-based corrosion inhibitor coating on carbon steel specimen displayed superior barrier properties to reduce corrosion rate up to 92.9% in the 3.5 wt% NaCl solution. GG/EP-coated carbon steel specimens make it more difficult for corrosive media to reach the substrate than uncoated. This demonstrates the stability and dissolving resistance of the surface coated with the GG/EP. Such development significantly strengthens the coating industries, which seek to prevent corrosion in various corrosive environments.

References

1. Cui, G., Bi, Z., Wang, S., Liu, J., Xing, X., Li, Z., Wang, B.: A comprehensive review on smart anticorrosive coatings. Prog. Org. Coat. (148), 105821 (2020)
2. Dariva, C.G., Gallio, A.F.: Corrosion Inhibitors–Principles, Mechanisms, and Applications. Developments in Corrosion Protection, INTECH (2014)
3. Zhiyong, L., Wenju, L., Wei, W., Cuiwei, D., Xiaoguang, L.: Failure analysis of leakage caused by perforation in an L415 steel gas pipeline. Case Stud. Eng. Fail. Anal. **9**, 63–70 (2017)
4. Zafar, F., Ghosa, l.A., Sharmin, E.: A review on cleaner production of polymeric and nanocomposite coatings based on waterborne polyurethane dispersions from seed oils. Prog. Org. Coat. (131), 259–275 (2019)
5. Wang, S., Wu, Y., Dai, J.: Making organic coatings greener: Renewable resource, solvent-free synthesis, UV curing, and repairability. Eur. Polym. J. (123), 109439 (2020)
6. Jing, L.C., Wang, T., Cao, W.W.: Water-based polyurethane composite anticorrosive barrier coating via enhanced dispersion of functionalized GO in the presence of acidified multi-walled carbon nanotubes. Prog. Org. Coat. (146), 105734 (2020)

Chapter 25
Assessment of Mechanical Behavior of Ti-Based Biomaterial in Prosthesis Application

Ganesh Kumar Sharma and Vikas Kukshal

Abstract Biomaterials encompass a broad range of tissue engineering methods that make it possible to produce materials that can be utilized to repair or replace damaged or missing parts of biological systems. The use of metallic biomaterials in medical implants is complemented by biocompatibility as a prominent aspect. Titanium (Ti) and its alloys are employed as promising implant materials due to their high strength-to-weight ratio and strong biocompatibility. The biocompatibility of Ti-based biomaterials can be improved by altering their mechanical and structural properties, which improves their interaction with the biological environment. Emerging research in advanced Ti-based alloy biomaterials is focused on creating implants with a low elastic modulus while still providing adequate mechanical strength. The major goal is to extend the biomaterial implant's lifespan or eliminate the need for revision surgery, hence improving the patient's quality of life. The present study focuses on Titanium based biomaterial developed with minor alloying elements such as Nb and Zr. Nb and Zr are non-toxic and mechanically sound materials for use in prosthetics. A computational strategy based on the finite element method is utilized to estimate the mechanical properties of the Ti–8Nb–12Zr alloy for biomedical applications. It is concluded that the Ti–Nb–Zr biomaterial with a dominant β phase microstructure exhibits a good combination of mechanical properties and biocompatibility for prosthesis application.

Keywords Titanium alloy · Biomaterial · Mechanical properties · Finite element method

G. K. Sharma · V. Kukshal (✉)
Department of Mechanical Engineering, National Institute of Technology Uttarakhand, Srinagar, Garhwal 246174, India
e-mail: vikaskukshal@nituk.ac.in

G. K. Sharma
Department of Mechanical Engineering, Moradabad Institute of Technology, Moradabad 244001, India

A. Tewari et al. (eds.), *Proceedings of the 3rd International Conference on Advances in Materials Processing: Challenges and Opportunities*, Springer Proceedings in Physics 293, https://doi.org/10.1007/978-981-99-1971-0_25

25.1 Introduction

Metal alloys play a significant role in modern medicine and are widely used in several medical sectors. However, due to the potential for negative consequences caused by freed ions in the surrounding tissues, only a few metals are appropriate for use in permanent implants. Biocompatible materials must be engineered to function under biological restraint and then adapt to various medical functions. A biomaterial is any substance or combination of components, whether natural or manufactured, that can be employed externally as a component of a system for morphological reconstruction and/or functional tissues and organs [1, 2].

Stainless Steel, Titanium, Magnesium and Cobalt-based alloys have exceptional qualities among all metals for biomedical applications. In addition to biocompatibility and osseointegration, Ti and Ti-based alloys also exhibit high wear and corrosion resistance as well as strength. There are numerous applications of pure titanium, titanium alloys and titanium matrix composites in various industries [3, 4]. Titanium materials are classified depending on their equilibrium state structure. In biomedical applications and alloys with the α, $\alpha + \beta$ and β structures are used in biomedical applications. Various studies have shown that the negative biological effects of metallic alloys can be reduced by altering their chemical makeup, microstructure and surface reactivity [5, 6].

The noble metals Tantalum (Ta), Niobium (Nb), Zirconium (Zr) and their alloys are quickly becoming desirable solutions for the majority of medical applications. Niobium exhibits a similar effect to Vanadium in stabilizing β phase in the Ti-Nb binary system, which is necessary for providing the α, β two-phase structure. Nb is a lightweight and cheaper material that has an elastic modulus close to the bones and superior corrosion resistance than pure titanium. Ti–6Al–7Nb alloy oxidized at 800 °C for 72 h had six times higher wear resistance than non-oxidized Ti–6Al–7Nb [7, 8]. Al and V are considered hazardous materials and have poor tribological characteristics when used over an extended period. Medical devices use the Zr–2.5Nb alloy to combat the problem of artificial joint sliding wear. Ti–Zr orthopaedic implant with increased mechanical performance and high load capacity 13–17% Zr. A self-consistent bcc phase atomic mobility database was produced by analysing the diffusion coefficients of the Ti–Mo–Zr system [9]. Ti–Nb–Ta–Zr alloy was changed by Fowler et al. by including antimicrobial Cu. The addition of a few percent of Cu has a considerable impact on microstructural changes, increasing the hardness of alloys [10, 11]. The mechanical behavior, such as strength, fatigue, fracture behavior and deformation, can be examined through the simulation method [12]. However, biomechanics simulation is challenging as the fabrics of material properties are very complex. Nevertheless, simulation was used successfully to calculate mechanical behavior and improve processing methods that increase biomaterials' lifespan [13, 14].

The qualities of the finished material are directly influenced by the alloying components; thus, it is important to choose the best percentage. The study discusses the fabrication of Ti-based alloy using elements like Nb and Zr evaluates the effects of the lements on physical-chemical and biological properties.

25.2 Materials and Methodology

This work is carried out in three stages; 3D Scanning of the prosthesis using InVesalius software, prosthesis modeling using CATIA and mechanical analysis of the prosthesis using ANSYS tool. The step-by-step methodology adopted in the study of femur bone prosthesis is as follows:

The objective of the present research work is to understand the various physical and mechanical properties of Titanium, Zirconium and Niobium alloy. The study of chemical, physical and mechanical properties of the metals used in the alloy can help with material selection along with the corrosion behavior, strength analysis, density, biocompatibility and the metal's ability to fuse with the bone and other tissues [15, 16]. Table 25.1 presents the physical and mechanical properties of the alloying materials.

The present study is mainly focused on the four parameters of the Hume-Rothery Rule (Atomic radius, Crystal structure, Electronegativity and Valence) required for the formation of the alloys. The proposed Titanium alloy-based metallic biomaterials are developed with minor alloying elements of 8% Nb and 12 % Zr.

The digital imaging and communications in medicine (DICOM) image of the prosthesis is used to build a three-dimensional model of a healthy femur. The geometry of the femur and material assignment are two variables that influence the meshing of the model. The flowchart for the design and file generation of the Prosthesis is shown in Fig. 25.1. To examine the stress distribution in the bone-implant interface, a hip-stem implant was made based on the typical geometry of the prosthesis. The three-dimensional model of the prosthesis is developed using CATIA software as shown in Fig. 25.2.

The entire assembly is made up of 8 nodes of tetrahedral pieces, with a maximum triangle edge of 0.5 mm and a shape quality threshold of 0.2. This tetrahedral volume mesh, which has 519,598 elements and 856,897 nodes, was used in basic mesh for the FEA analysis. The non-linear geometry of hip stem assemblies requires the study of the FEA using the dynamic implicit approach.

Table 25.1 Physical and mechanical properties of alloying elements [17–20]

Alloying elements	Density (g/cm^3)	Atomic weight (g/mol)	Tensile strength (MPa)	Modulus of elasticity (GPa)	Shear modulus (GPa)	Poisson ratio
Ti	4.5	47.8	220	116	43	0.34
Nb	8.3	92.96	276	106	38	0.40
Zr	6.3	91.2	230	94	35	0.30

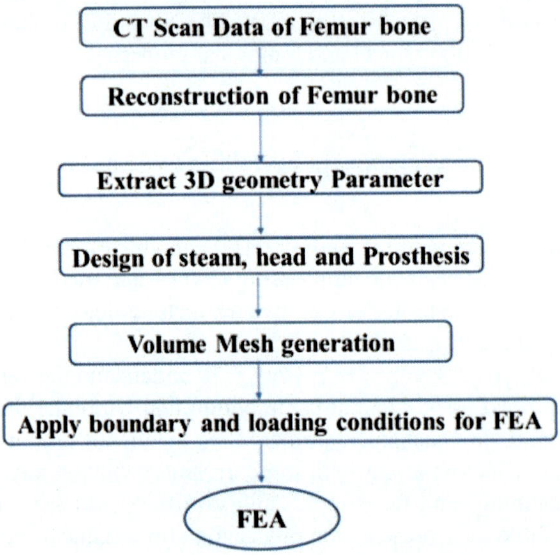

Fig. 25.1 Flow Chart for design and file generation of Prosthesis (with the help of the quasi-static, the standard load variation over time has been adjusted to ramp linearly over a defined step)

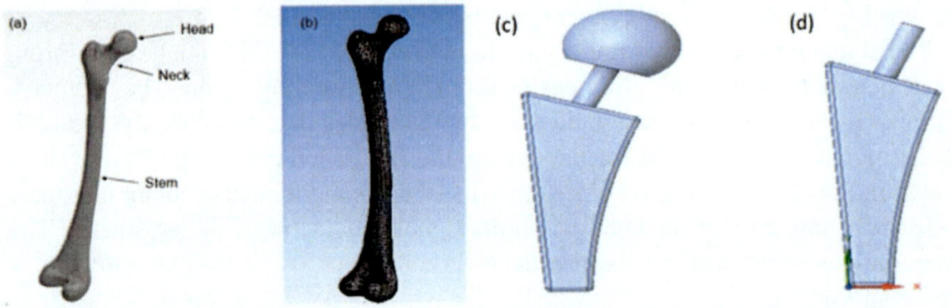

Fig. 25.2 **a** CAD model of femur bone **b** 3D volume meshed of femur bone **c** CAD model of assembled prosthesis with head, neck and stem **d** CAD model of bone stem part

25.3 Results and Discussion

The deformation, stress and strain results of Ti–8Nb–12Zr alloys in the direction of the applied load are estimated by ANSYS Workbench. The analysis of the model observed that total deformation is 0.612 mm, minimum and maximum stress value is 11.4 MPa and 660.5 MPa respectively and elastic strain is 3.95 mm/mm. The analysis of femur bone observed that the total deformation is 9.98 mm, minimum stress value is 0.0028 MPa, maximum stress value is 5.78 MPa, and Elastic Strain 0.00285 mm/mm are shown in Fig. 25.3. The result of this study is presented in Table 25.2.

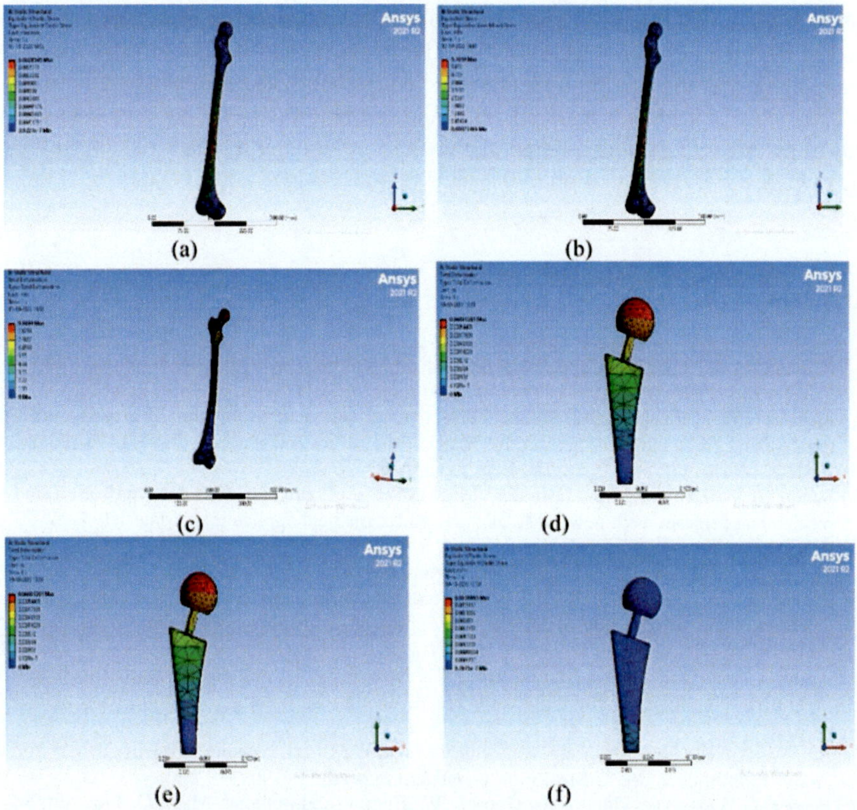

Fig. 25.3 **a** Equivalent elastic strain [mm/mm] of femur bone **b** equivalent (Von-Mises) stress [MPa] of femur bone **c** total deformation [mm] of femur bone **d** total deformation [mm] of prosthesis (CAD model of femur bone with head, neck and stem) **e** equivalent elastic strain [m/m] of prosthesis (CAD model of femur bone with head, neck and stem) **f** equivalent elastic strain [Pa] of prosthesis (CAD model of femur bone with head, neck and stem)

Table 25.2 Comparison between Femur bone and prosthesis results of Mechanical Properties.

Studied cases	Total deformation (mm)	Equivalent stress (MPa)	Elastic strain (mm/mm)
Femur bone	9.98	5.71	0.00285
Prosthesis	0.612	660.5	3.95

25.4 Conclusion

The stress produced in the hip stem and femur bone under the loading conditions is examined in this study using finite element analysis. During the design of a prosthesis, neck diameter plays a vital role in optimization. The results show that the model is safe with optimum design. The maximum von mises stresses are observed near the head of the prosthesis and neck. The prosthesis generates the most stress with the stem being under the safe limit and having good strength and mechanical qualities. Ti–8Nb–12Zr is a perfect material for femur implant due to its good mechanical properties. As a result, it is inferred that the neck region has a higher risk of failure due to high stress concentration in the same region.

References

1. Niinomi, M.: Mechanical properties of biomedical titanium alloys. Mater. Sci. Eng.: A **243**(1–2), 231–236 (1998), ISSN 0921-5093. https://doi.org/10.1016/S0921-5093(97)00806-X
2. Sharma, G.K., Kukshal, V.: Assessment of mechanical behaviour of functionally graded biomaterials using finite element method. In: Mater. Today: Proc. **62**, Part 14, 7487–7493 (2022), ISSN 2214-7853. https://doi.org/10.1016/j.matpr.2022.03.513.
3. Sandu, A.V., Baltatu, M.S., Nabialek, M., Savin, A., Vizureanu, P.: Characterization and mechanical proprieties of new TiMo alloys used for medical applications. Materials **12**, 2973 (2019)
4. Sharma, G.K., Kukshal, V.: Mechanical behaviour and finite element analysis of biomaterials: a review. In: Krishnapillai, S., R., V., Ha, S.K. (eds.) Composite Materials for Extreme Loading . Lecture Notes in Mechanical Engineering. Springer, Singapore (2022). https://doi.org/10.1007/978-981-16-4138-1_26
5. Aniołek, Krzysztof, Kupka, Marian, Łuczuk, Marta, Barylski, Adrian: Isothermal oxidation of Ti-6Al-7Nb alloy. Vacuum. **114**, 114–118 (2015). https://doi.org/10.1016/j.vacuum.2015.01.016
6. Bai, W., Xu, G., Tan, M., Yang, Z., Zeng, L., Wu, D., Liu, L., Zhang, L.: Diffusivities and atomic mobilities in bcc Ti-Mo-Zr alloys. Materials **2018**, 11 (1909)
7. Fowler, L., Janse Van Vuuren, A.; Goosen, W., Engqvist, H., Öhman-Mägi, C., Norgren, S.: Investigation of copper alloying in a TNTZ-Cux alloy. Materials **12**, 3691 (2019)
8. Okabe, T., Ohkubo, C., Watanabe, I., Okuno, O., Takada, Y.: JOM, 24–29 (1998)
9. Miyamoto, Y., Kaysser, W.A., Rabin, B.H., Kawasaki, A., Ford, R.G. (eds.): Kluwer Academic Publishers, Boston, USA (1999)
10. Watari, F., Takahashi, H.: Trans. 8th Symp. Apatite, 'Jap. Assoc. Apatite Sci., Tokyo Med. Dent, Univ., Tokyo, pp. 25–26 (1992)
11. Zadpoor, A.A., Malda, J.: Additive manufacturing of biomaterials, tissues, and organs. Ann. Biomed. Eng. **45**, 1–11 (2017)
12. Sharma G.K., Kukshal V., Shekhawat, D., Patnaik, A.: Fabrication and Characterization of Metallic Biomaterials in Medical Applications Advanced Materials and Manufacturing Processes, 1st edn. CRC Press (2021). https://doi.org/10.1201/9781003093213
13. Murr, L., Gaytan, S., Medina, F., Lopez, H., Martinez, E., Machado, B., Hernandez, D., Martinez, L., Lopez, M., Wicker, R.: Next-generation biomedical implants using additive manufacturing of complex, cellular and functional mesh arrays. Philos. Trans. R. Soc. London A: Math. Phys. Eng. Sci. **368**, 1999–2032 (2010)
14. Zadpoor, A.A.: Mechanical performance of additively manufactured meta biomaterials. Acta Biomater. **85**, 41–59 (2019)
15. Tan, X., Tan, Y., Chow, C., Tor, S., Yeong, W.: Metallic powder-bed based 3D printing of cellular scaffolds for orthopaedic implants: a state-of-the-art review on manufacturing, topological design, mechanical properties and biocompatibility. Mater. Sci. Eng., C **76**, 1328–1343 (2017). https://doi.org/10.1016/j.msec.2017.02.094
16. Sing, S.L., An, J., Yeong, W.Y., Wiria, F.E.: Laser and electron-beam powder-bed additive manufacturing of metallic implants: a review on processes, materials and designs. J. Orth. Res. **34**, 369–385 (2016). https://doi.org/10.1002/jor.23075

17. Wang, X., Li, Y., Xiong, J., Hodgson, P.D., Wen, C.: Porous TiNbZr alloy scaffolds for biomedical applications. Acta Biomater. **5**(9), 3616–3624 (2009), ISSN 1742-7061. https://doi.org/10.1016/j.actbio.2009.06.002
18. Masahashi, N., Mizukoshi, Y., Semboshi, S., Ohmura, K., Hanada, S.: Photo-induced properties of anodic oxide films on Ti6Al4V. Thin Solid Films **520**(15), 4956–4964 (2012)
19. Bolzoni, L., Ruiz-Navas, E.M., Neubauer, E., Gordo, E.: Mechanical properties and microstructural evolution of vacuum hot-pressed titanium and Ti-6Al-7Nb alloy. J. Mech. Behav. Biomed. Mater. **9**, 91–99 (2012)
20. Ho, W.F., Chen, W.K., Wu, S.C., et al.: Structure, mechanical properties, and grindability of dental Ti–Zr alloys. J. Mater. Sci: Mater. Med. **19**, 3179–3186 (2008). https://doi.org/10.1007/s10856-008-3454-x

Chapter 26
Challenges Associated with 3D Printing of Poly-Ether-Ether Ketone (PEEK) Components

Gaurav Sharma, Amol Vuppuluri, and Kurra Suresh

Abstract The fused deposition modeling (FDM) technique has accrued widespread recognition in both academia and industries owing to its outstanding capability of printing thermoplastics with product customization, cost effectiveness, reliability, ease of operation and handling. Moreover, 3D printing technology facilitates the printing of complex geometries with dimensional stability and good resolution. However, a profound knowledge of the choice of printing parameters and machine settings is imperative for printing high-temperature thermoplastics like PEEK because owing to non-uniform heat distribution during the 3D printing, PEEK is prone to warpage and differential crystallinity. Moreover, other FDM-based defects could also arise on account of situations involving excessive moisture intake by a filament or maintaining improper bed leveling. This research article briefly discusses about the determinant factors responsible for defect generation in 3D-printed PEEK components and suggests their amends based on our own experiential learning during PEEK 3D printing. This study would corroborate various parameter setting necessary for eliminating irregularities and obtaining defect-free printed PEEK components with superb mechanical and fracture behaviors.

Keywords Fused deposition modeling (FDM) · Poly-ether-ether-ketone (PEEK) · Warpage

26.1 Introduction

Commercialized by Stratasys © in the early 1990s, FDM is a 3D printing technology involving layer-by-layer deposition of engineering thermoplastics in a semi-solid state, according to the intended geometry [1]. In the league of thermoplastics, PEEK is a high-performance semi-crystalline thermoplastic with outstanding mechanical

G. Sharma (✉) · A. Vuppuluri · K. Suresh
Department of Mechanical Engineering, Birla Institute of Technology and Science (BITS) Pilani, Hyderabad Campus, Secunderabad 500078, Telengana, India
e-mail: p20200464@hyderabad.bits-pilani.ac.in

A. Tewari et al. (eds.), *Proceedings of the 3rd International Conference on Advances in Materials Processing: Challenges and Opportunities*, Springer Proceedings in Physics 293, https://doi.org/10.1007/978-981-99-1971-0_26

and thermal properties, chemical and impact resistance, biocompatibility, recyclability and radiolucency. These stellar properties have intrigued a lot of enthrallments in numerous critical applications involving biomedical, automotive, aerospace, electrical and coatings. However, the high melting temperature (~343 °C) and semi-crystalline nature of PEEK hinders its FDM-based processing because of excessive thermal stresses generated due to uneven thermal gradients during 3D printing which further leads to development of defects like warpage, varying crystallinities, insufficient bonding between deposited layers, inaccurate dimensions of final product etc. [2]. Nevertheless, there exists a very strong conformity between the ultimate property and printing prerequisites (i.e. part parameters and machine settings) selected for 3D printing and henceforth, this research article is an initiative to identify the abatement techniques utilized by us for sabotaging the defects arising from nonuniform thermal gradients induced during 3D printing of PEEK, thereby ultimately resulting into 3D printed PEEK components with excellent structural integrity and reliability.

26.2 Material and Method

The FDM-based 3D printing of double edge notch tensile (DENT) specimens was accomplished using INTAMSYS FUNMAT HT printer whereas the PEEK filament (diameter = 1.75 mm) was procured from Loadstar Innovations Pvt. Ltd., Bangalore. The computer aided drawing (CAD) model was converted into standard tessellation language (STL) format, followed by slicing using INTAMSUITE 3.4.1 software. The printing parameters chosen are shown in Table 26.1.

Table 26.1 FDM parameters opted for 3D printing PEEK

Parameter name	Parameter value
Build plate temperature (°C)	130
Extrusion temperature (°C)	420
Chamber temperature (°C)	80
Fan speed (%)	50
Infill density (%)	100
Deposition width (mm)	0.4
Print speed (mm/s)	20
Layer thickness (mm)	0.12

26.3 Challenges in 3D Printing PEEK

26.3.1 Warpage

The non-homogeneous thermal gradient existing during 3D printing of semi-crystalline PEEK during the cooling down process from extrusion temperature to ambient temperature leads to the core of the sample cooling at a slower rate as compared to the outer surface [3]. We recommend maintaining a high chamber temperature in order to decrease the cooling rate as well as for achieving more uniform heat distribution within build part, thereby leading to controlled crystal growth and elimination of residual stresses. Since the warpage is observed along the edges and bottom layers of printed component, it is recommended to maintain the elevated build plate temperature for ensuring excellent adhesion between the platform and depositing material and obtain warp-free establishment by abating quick solidification of initial few extruded layers. Both elevated chamber temperature and platform temperature also ameliorate the inter-layer bonding strength, which further has positive influence over mechanical and fracture characteristics of the printed components.

We also recommend using a washable glue stick for creating a glutinous surface on the build plate before commencing the 3D printing. Cooling fan ensures that the material extruding out of the nozzle is frozen at the earliest in order to avoid any erroneous movement or deformations which are not conforming with GCODE. However, it is also of paramount importance to avoid fast solidification of initial layers for preventing warpage (see Fig. 26.1) and henceforth, we recommend switching off the cooling fan for the initial few layers and again turning it on for the subsequent layers.

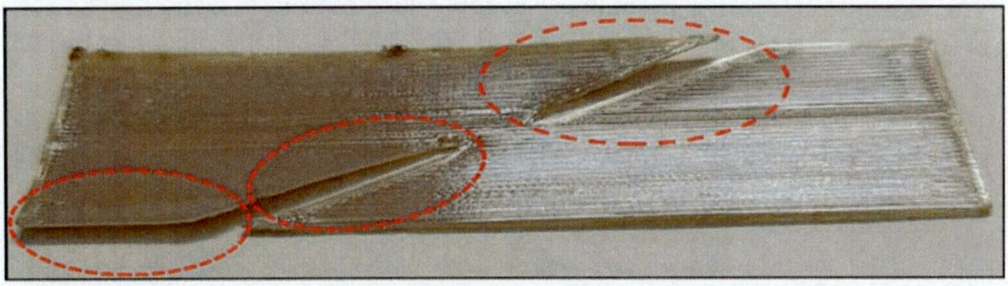

Fig. 26.1 Warpage in 3D printed PEEK because of uneven cooling rates between previously deposited and freshly deposited layer (depicted by red circles)

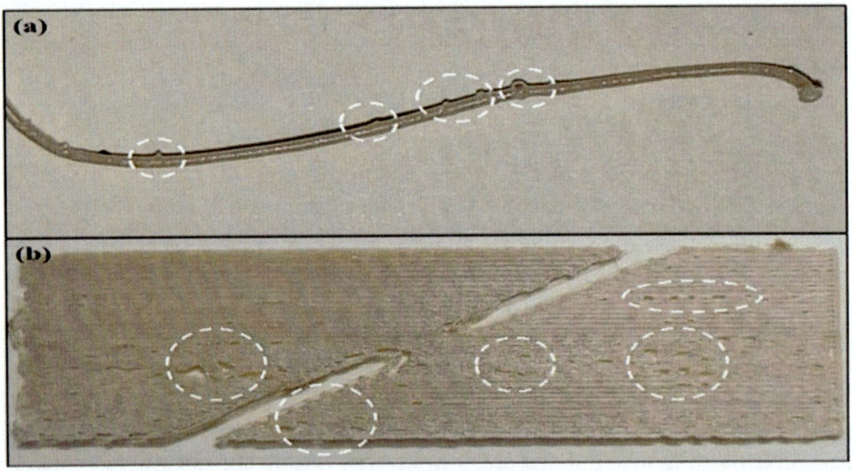

Fig. 26.2 **a** Blobs and **b** Pockmarks present in PEEK wire filament and final 3D printed specimen, respectively because of inappropriate filament storage technique (depicted by white circles)

26.3.2 Moisture Intake by Filament

All thermoplastic filaments have a tendency of moisture absorption However, the presence of moisture imparts additional anisotropy into already anisotropic 3D printed thermoplastics by negatively impacting the inter-layer bonding strength [4]. Moisture engrossment leads to filament swelling and alterations in glass transition temperature (T_g), which further impacts the viscosity of the extruded material and causes extrusion jam and delays, formation of blobs in the filament, stringy filament and small pockmarks in the printed specimen (see Fig. 26.2).

The moisture intake by the filament at times leads to plasticization effect [5], resulting in increasing the failure strain and reduction in tensile modulus, tensile strength as well as T_g. The best practices for removing moisture from thermoplastic filaments is pre-heating at an elevated temperature for an explicit amount of time, as specified by the manufacturers of the FDM printer. It should be followed by allowing the filament spool to cool down inside the hot air oven and then transferring it to a dry box containing silica gel packets.

26.3.3 Low Crystallinity Regions

The 3D-printed PEEK specimens contained both crystalline regions (opaque) and amorphous regions (transparent). The characteristics of the amorphous region involve transparency, high elasticity, less chemical resistance, and low density, whereas its crystalline counterpart indicates higher elastic modulus, tensile strength, and stress-cracking resistance when compared to the amorphous region [6]. These low-density

Fig. 26.3 Crack propagation along the low crystallinity transparent region during mode I loading of 3D printed PEEK having 45° print orientation

regions at times act as sites of crack initiation and propagation. For example, during the toughness evaluation of DENT specimens, where the crack is supposed to propagate in the region joining the crack tips, it is observed that the presence of the amorphous region ahead of the crack resulted in the crack propagation at a certain orientation (see Fig. 26.3).

Yang et al. [7] recommended performing post-printing thermal processing (post-annealing) and maintaining high chamber temperature for improving crystallinity and obtaining better mechanical response. Other than maintaining high chamber temperature, we also suggest allowing the 3D printed component to cool within the chamber for at least 0.5 h before removing it.

26.3.4 Improper Bed Leveling

Obtaining a suitable first layer is quintessential for ensuring a consistent and good quality print with legitimate dimensional accuracy. However, an inconsistent gap between platform and nozzle would lead to nozzle blockage, grotesque first layer, nozzle scratching against the build platform, filament not extruding at specific locations of print, extruded filament not adhering to build plate and filament geometry varying across build platform. Therefore, we recommend performing manual bed leveling, automatic bed leveling as well as level sensor calibration using feeler gauges.

26.4 Conclusion

This research article presents guidelines regarding printing conditions and machine settings that have to be taken care of before and during the 3D printing of PEEK for obtaining defect-free components with excellent mechanical and fracture characteristics. We exhort the elevated chamber and platform temperatures along with washable glue sticks as an adhesive for regulating warpage and eventually obtaining a dimensionally veracious component. Moreover, adjusting the status (ON / OFF) of the cooling fan between the initial and subsequent layers is also indicated. Preheating the PEEK filament and storing it in a dry box containing silica gel sachets is recommended for moisture elimination and ensuring the absence of pockmarks in printed components. The relationship between crystallinity and mechanical properties is established and post-annealing heat treatment has been recommended as an appropriate route for increasing crystallinity. Finally, various indicators of improper bed level along with manual and automatic leveling and lever sensor calibration using feeler gauges have also been discussed. Henceforth, the contents of this investigation would be certainly beneficial for the researchers and professionals attempting to print semi-crystalline thermoplastics with excellent quality and superior properties.

References

1. Penumakala, P.K., Santo, J., Thomas, A.: A critical review on the fused deposition modeling of thermoplastic polymer composites. Compos. B **201**, 108336 (2020)
2. Sharma, G., Vuppuluri, A., Suresh, K.: Essential work of fracture studies of 3D printed PEEK (Poly-ether-ether-ketone) polymer. Eng. Fract. Mech. **271**, 108656 (2022)
3. Samy, A.A., Golbang, A., Harkin-Jones, E., Archer, E., McIlhagger, A.: Prediction of part distortion infused Deposition Modeling (FDM) of semi-crystalline polymers via COSMOL: effect of printing conditions. CIRP J. Manuf. Sci. Technol. **33**, 443–453 (2021)
4. Wichniarek, R., Hamrol, A., Kuczko, W., Górski, F., Rogalewicz, M.: ABS filament moisture compensation possibilities in the FDM process. CIRP J. Manuf. Sci. Technol. **35**, 550–559 (2021)
5. Zaldivar, R.J., Mclouth, T.D., Ferrelli, G.L., Patel, D.N., Hopkins, A.R., Witkin, D.: Effect of initial filament moisture content on the microstructure and mechanical performance of ULTEM® 9085 3D printed parts. Addit. Manuf. **24**, 457–466 (2018)
6. Wang, R., Cheng, K., Advincula, R.C., Chen, Q.: On the thermal processing and mechanical properties of 3D-printed polyether ether ketone. MRS Commun. **9**, 1046–1052 (2019)
7. Yang, C., Tian, X., Li, D., Cao, Y., Zhao, F., Shi, C.: Influence of thermal processing conditions in 3D printing on the crystallinity and mechanical properties of PEEK material. J. Mater. Process. Technol. **248**, 1–7 (2017)

Chapter 27
Effect of Functionalization of Carbon Nanotubes on Mechanical Behavior of Ultra High Molecular Weight Polyethylene Matrix Composite

Krishnakant Phand, Bhavya Jain, and Debrupa Lahiri

Abstract Ultra-High Molecular-Weight Polyethylene (UHMWPE) is one of the most commonly used polymers in research fields because of its superior toughness, impact strength, low friction and sliding wear resistance. In terms of mechanical, thermal and electrical properties, carbon nanotubes (CNTs) are an excellent choice for reinforcement. Being chemically inert in nature, CNTs does not readily bond with polymers. Hence, there is scope of improvement in interfacial bonding by covalent functionalization of CNTs. This study aims to investigate the effect of functionalized CNT (f-CNT) reinforcement on the mechanical, thermal and structural properties of UHMWPE composites. Multi-walled carbon nanotubes (MWCNTs) are acid functionalized and composites with different CNT and (f-CNT) loading proportions are fabricated using compression molding followed by curing. With addition of (1 wt. %) f-CNT a 75% increase in crystallinity is observed. The yield strength and elastic modulus nearly doubled for 1wt% f-CNT reinforced composites. Similarly, the storage modulus is found to be increased nearly 6 times. Overall, f-CNT reinforced composites exhibited better structural, mechanical and thermal behavior because of higher interfacial interaction between f-CNTs and UHMWPE, and uniform dispersion of f-CNTs in the composite matrix.

Keywords Functionalized CNT · Mechanical behavior · Thermal behavior

K. Phand · D. Lahiri (✉)
Indian Institute of Technology Roorkee, Roorkee 247667, India
e-mail: debrupa.lahiri@mt.iitr.ac.in

B. Jain
Galgotias College of Engineering and Technology, Greater Noida 201310, India

A. Tewari et al. (eds.), *Proceedings of the 3rd International Conference on Advances in Materials Processing: Challenges and Opportunities*, Springer Proceedings in Physics 293, https://doi.org/10.1007/978-981-99-1971-0_27

27.1 Introduction

CNT exhibits a unique graphitic structure, excellent electrical, thermal, and optical properties and has attracted interest in biomedical, military, and other different industries [7]. On the other hand, UHMWPE possesses high toughness and wear resistance, and is thus used in a variety of research and application fields [1]. Huang [3] discussed the toxicity and high agglomeration rate of CNTs that may hinder their biomedical applications in neural, cardiac, and bone engineering applications. Further, [6] fabricated UV and gamma irradiated nano composites of UHMWPE with Alumina, hydroxyapatite, and CNTs using compression molding. The UV and gamma-irradiated composites showed lower hardness, due to recrystallization and high agglomeration. Maksimkin [4] fabricate the composites of UHMWPE reinforced MWCNTs using hot pressing and found that the Composite with 1% MWCNT showed four times increase in ultimate strength and a significant increase in yield strength. To the best of the author's knowledge, none of the reported literature has presented the effect of functionalization on UHMWPE/CNT composites, particularly on the mechanical properties. In this work, CNT and f-CNT reinforced UHMWPE composites are prepared and analysed.

The primary theme of this work is to understand mechanical behaviour of UHMWPE after reinforcing f-CNT. In this regard, first covalent functionalization of MWCNT is performed, followed by preparation of UHMWPE composites reinforced with different loading proportions of CNT/f-CNT (0.1, 0.5, 1 wt. %) using compression molding. Further, the effect on mechanical and thermal properties of formed composites is investigated by dynamic mechanical analysis (DMA), compression testing and differential scanning calorimetry (DSC).

27.2 Experimental

27.2.1 Materials Used

MWCNTs (outer diameter 10–12 nm, length 8–12 μm, purity >95% and density 1.8 g/cm^3) used in this study were procured from C-Nano Technology Limited, Beijing, China. UHMWPE (average particle size 120 μm, density 0.96 g/cm^3) were obtained from JP Polymers Mumbai, India. Other chemicals like sulfuric acid (H_2SO_4), and nitric acid (HNO_3), are used as received from commercial sources.

27.2.2 Experimental

Unmodified CNT (1.0 g) were dispersed in an acid solution (60 mL of HNO_3 and 20 mL of H_2SO_4) at 90 °C and refluxed for 24 h, followed by centrifugation at

6000 rpm, until the pH value approached neutral (6–7). The resulting oxidized CNTs were dried at 85 °C for 12 h. The mixture of CNT and acetone was ultrasonicated for 60 m using a probe sonicator at 50 kHz. UHMWPE powder was added slowly and sonicated for another 30 m. The formed composite powder was then heated at 80 °C for 30 m. Further, disk shaped pellets of 30 mm diameter and 3 mm thickness and cylindrical specimen of 10 mm diameter and 25 mm length were formed using compression molding at 300 MPa, followed by curing at 180 °C for 60 m in case of UHMWPE pellets and for 90 m for CNT-UHMWPE pellets as CNTs lead to increase in melting temperature.

27.2.3 Characterization

Fourier Transform Infra-Red Spectroscopy (FTIR) was performed using PerkinElmer Spectrum Two FT-IR Spectrometer for chemical analysis of CNTs. The Morphological analysis of the composites was carried out using Field Emission Scanning Electron Microscopy (FE-SEM) on Carl Zeiss Ultra Plus. Thermal Analysis was done using DSC at a temperature of up to 500 °C at a rate of 5 °C/min in NETZSCH DSC 204F1 Phoenix, and the degree of crystallinity was calculated. Compression testing done for 10 mm diameter and 15 mm long cylindrical samples on Instron 5982 Testing machine according to the ASTM D695 standards. Furthermore, viscoelastic properties were determined using PerkinElmer DMA 800.

27.3 Result and Discussion

27.3.1 Chemical and Morphological Analysis

Figure 27.1(a) presents the FTIR spectra of functionalized (f-CNT) and pristine nanotubes (CNT). For pristine CNTs, the peak is obtained at 3448 cm^{-1}, which is attributed to (-OH) group, resulted from atmospheric moisture. For f-CNTs, peaks at 1720 cm^{-1} and 3423 cm^{-1} are obtained, which are the signatures the carbonyl (C=O) and hydroxyl stretching vibration of the carboxylic group (COOH), respectively. This observations prove successful functionalization of CNTs through acid treatment, which is supposed to help in better bonding of them in polymer matrix. Moreover, Fig. 27.1(b) (iii) and (iv) show that CNTs are well dispersed in UHMWPE matrix, compared to Fig. 27.1(b) (i) and (ii) because of reduced surface energy of CNTs, as well as, the possibility of chemical bonding due to functionalization.

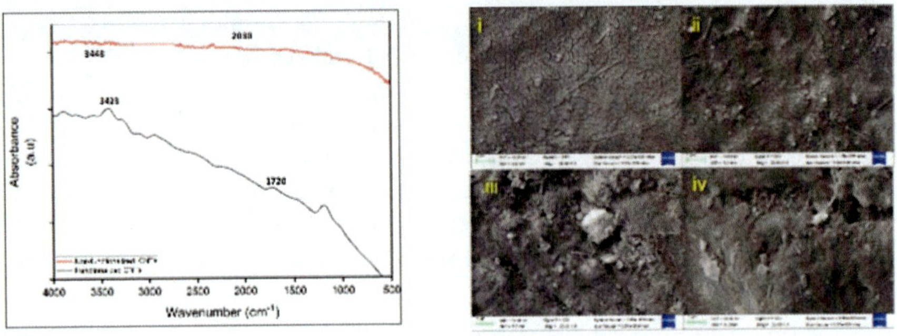

(a) (b)

Fig. 27.1 **a** FTIR spectra of CNT/f-CNTs, **b** SEM images of (i) CNT 0.5wt%, (ii) CNT 1wt%, (iii) f-CNT 0.5wt% and (iv) f-CNT 1wt% reinforced composites

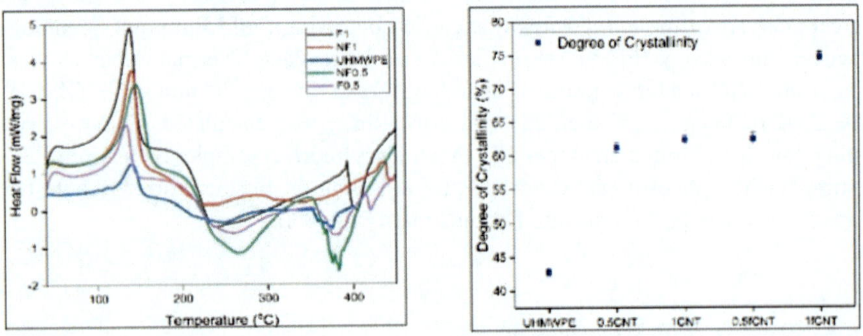

(a) (b)

Fig. 27.2 **a** DSC plots of composites, **b** Plot of calculated Crystallinity

27.3.2 Thermal Behavior

With increase in CNT loading proportion, crystallinity of composites is found to be increased, as CNTs act as nucleating sites in the UHMWPE matrix. Uniform dispersion of CNTs increases the number of nucleation sites thus increasing the crystalline lamellar structures within the matrix [2]. As shown in Fig. 27.2(b), With addition of 1wt. % f-CNTs to UHMWPE, 75% and 20% increase in crystallinity is observed compared to pure UHMWPE and 1wt. % CNT/UHMWPE specimen, respectively. Crystallinity calculated from the expression $Xc = \frac{\Delta Hc}{\Delta H100} x100$. ΔH_c is the enthalpy of crystallization of the specimen calculated by integrating the area under the peak. ΔH_{100} is the enthalpy of crystallization for pure crystalline samples. It is 289 J/g for UHMWPE.

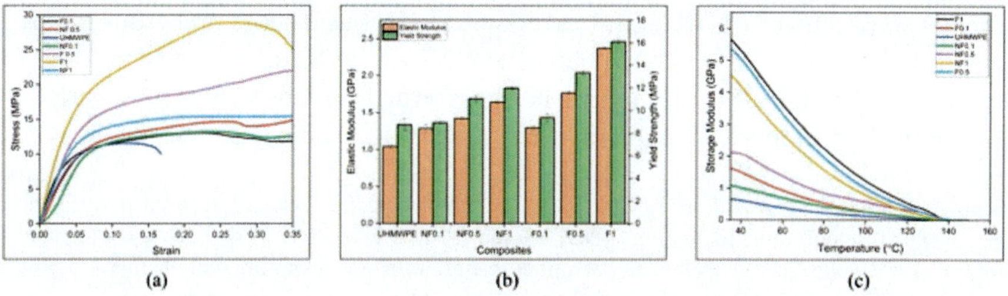

Fig. 27.3 Plot of: **a** Compression curves of composites **b** Combined plot for Yield Strength and Elastic Modulus, **c** Storage modulus curves of composites

27.3.3 Mechanical Behavior

Uniform dispersion of CNTs in UHMWPE matrix restricts the movement of polymer molecular chains under external loading as CNTs act as second phase particles. This hinders the distortion of molecular chains thereby increasing the strength and modulus of the composites. Moreover CNTs when uniformly distributed, are believed to form a bridged network structure throughout the entire matrix and aid in effective load transfer [2]. Figure 27.3(b) reveals nearly twofold improvement in elastic modulus (2.3 GPa) and yield strength (16 MPa) for 1wt. % f-CNT/UHMWPE composite, compared to that of pure UHMWPE (1 GPa and 9 MPa respectively). The values for the elastic modulus are fairly higher than those reported [5] for composites fabricated in a similar manner wherein maximum value of the Elastic modulus (1.4 GPa) was recorded for 0.1wt% loading proportion of non-functionalized CNTs. A remarkable 30% higher yield strength in 1wt. % is recorded in f-CNT composites, than 1wt. % CNT/UHMWPE composites. A sixfold rise in storage modulus for 1wt. % f-CNT/UHMWPE composites is also observed as compared to pure UHMWPE Fig. 27.3(c). Storage modulus of 1wt. % f-CNT/UHMWPE composite is 23% greater than that of 1wt. % CNT/UHMWPE composite.

27.4 Conclusion

In this work, CNT and f-CNT reinforced UHMWPE composites with different loading proportions (0.1, 0.5 and 1wt% each) were fabricated and the effect of functionalization was investigated on the mechanical, thermal and structural properties of the composites. CNTs were more uniformly distributed in UHMWPE in the case of f-CNTs owing to the fact that functionalization reduces the surface energy of the CNTs. Crystallinity was improved by 75% with loading proportion of 1wt% f-CNTs. Yield strength and Elastic modulus almost doubled compared to that of pure UHMWPE with addition of 1wt% f-CNTs. The viscoelastic properties also improved with reinforcement of CNTs. Nearly a six-fold rise in storage modulus was observed.

It can be inferred that f-CNTs reinforced composites showed better properties for the same loading proportions due to the fact that f-CNTs were better dispersed and had a better bonding with UHMWPE matrix. Further formed composites can have various applications in orthopedic implants as well as structural and defense components.

Acknowledgements The authors would like to acknowledge the financial assistance from the Department of Science and Technology (DST), India [SB/SO/HS/ 138/2013]. The authors extend their ssincere thanks to the technical personnel maintaining the experimental facilities at the Department of Metallurgical and Materials Engineering, IIT Roorkee.

References

1. Baena, J.C.: Wear performance of UHMWPE and reinforced UHMWPE composites in arthroplasty applications: a review. Lubricants **3**, 413–436 (2015)
2. Gupta, T.K.: Self-sensing and mechanical performance of CNT/GNP/UHMWPE biocompatible nanocomposites. J. Mater. Sci. **53**(11), 7939–7952 (2016)
3. Huang, B.: Carbon nanotubes and their polymeric composites: the applications in tissue engineering. Biomanuf. Rev. **5**(3) (2020)
4. Maksimkin: Ultra-high molecular weight polyethylene reinforced with multi-walled carbon nanotubes: fabrication method and properties. J. Alloys Compounds **536**, S538–S540 (2012)
5. Manoj, K.R.: Effects of carbon nanotube aspect ratio on strengthening and tribological behavior of ultra high molecular weight polyethylene composite. Compos. A Appl. Sci. Manuf. **76**, 62–72 (2015)
6. Nayak, C.: Radiation-induced effects on micro-scratch of ultra high molecular weight polyethylene biocomposites. J. Market. Res. **11**, 2277–2293 (2021)
7. Puértolas, J.A.: Evaluation of carbon nanotubes and graphene as reinforcements for UHMWPE-based composites in arthroplastic applications: a review. J. Mech. Behav. Biomed. Mater. **39**, 129–145 (2012)

Chapter 28
Influence of Different Precipitating Agents on the Synthesis of NiMn-LDHs Based Cathode Materials for High Performance Hybrid Devices

Megha Goyal and Tapas Kumar Mandal

Abstract Improving the capacity of transition metal hydroxide-based cathode materials is vital for creating hybrid devices with excellent electrochemical performance. Herein, we report the synthesis of NiMn layered double hydroxides (LDHs) by hydrothermal method, using Hexamethylenetetramine (HMTA) and urea as precipitating agents, and are named NiMn-LDH@HMTA and NiMn-LDH@urea, respectively. The compounds are phase pure, as evidenced by P-XRD analyses and crystallize in α-Ni $(OH)_2$ structure, while flower-like microstructure is confirmed by FESEM analysis. Electrochemical characterizations (CV, GCD and EIS) affirm the battery type charge storage mechanism for both samples and high specific capacitance/capacity of 1685 Fg^{-1} (675 Cg^{-1}) for NiMn-LDH@HMTA as compared to 1218 Fg^{-1} (487 Cg^{-1}) for NiMn-LDH@urea at 1 Ag^{-1}. A low solution resistance ($R_s = 0.47\ \Omega$) and charge transfer resistance ($R_{ct} = 7.5\ \Omega$) for NiMn LDH@HMTA results in high-capacity retention (82%) over 5000 GCD cycles at 10 Ag^{-1} confirming its superiority over NiMn-LDH@urea and potential to be used as a cathode material for practical energy storage devices.

Keywords LDHs · Hydrothermal synthesis · Energy storage

28.1 Introduction

Hybrid energy storage system (HESS) is one of the best alternatives to a traditional storage system such as batteries and supercapacitors. HESS consisting of one battery type electrode and another supercapacitor type electrode, offers merits of both the storage systems such as, high capacitance, cyclability and high energy and power

M. Goyal · T. K. Mandal (✉)
Centre for Nanotechnology, Indian Institute of Technology Roorkee, Roorkee 247667, India
e-mail: tapas.mandal@cy.iitr.ac.in

T. K. Mandal
Department of Chemistry, Indian Institute of Technology Roorkee, Roorkee 247667, India

A. Tewari et al. (eds.), *Proceedings of the 3rd International Conference on Advances in Materials Processing: Challenges and Opportunities*, Springer Proceedings in Physics 293, https://doi.org/10.1007/978-981-99-1971-0_28

density [1]. To get optimum performance from HESS, the supercapacitor material should be well matched with the battery type material in terms of capacity and mass loading [2]. To accomplish this, compatible battery-type cathode materials are in urgent need. Among all reported battery-type electrode materials, nickel-based oxides and double hydroxides are promising [3] due to their low cost, high abundance, low toxicity, high stability and multiple redox states. In the comparison of oxides, layered double hydroxides (LDHs) offer several advantages due to their exclusive layered structure, which helps to produce a large number of ion channels, high surface area and high electrical conductivity [4], leading to better electrochemical performance.

In this work, the effect of precipitating agents, such as, Hexamethylenetetramine (HMTA) and urea on the crystallite size, microstructure and electrochemical performance of NiMn-LDHs is investigated. A high specific capacitance of 1685 Fg^{-1} (657 Cg^{-1}) at 1 Ag^{-1} with 82% capacity retention over 5000 GCD cycles is obtained for NiMn-LDH@HMTA.

28.2 Experimental

28.2.1 Synthesis

The synthesis of NiMn-LDHs was carried out by a one-pot hydrothermal method. For the synthesis, first $Ni(NO_3)_2.6H_2O$, $Mn(NO_3)_2.H_2O$ and HMTA (in 1:1:0.7 mmol) and $Ni(NO_3)_2.6H_2O$, $Mn(NO_3)_2.H_2O$ and urea (in 1:1:6 mmol) was dissolved in 40 ml distilled water in two separate beakers. Then, the resulting solutions were magnetically stirred for 15 min to obtain uniform suspensions. The obtained suspensions were transferred into Teflon-lined autoclaves (50 ml capacity) and heated at 160 °C for 2 h and 90 °C for 5 h, for NiMn-LDH@HMTA and NiMn-LDH@urea, respectively. The obtained compounds were washed thoroughly with distilled water followed by absolute ethanol and dried in an air oven at 60 °C for overnight to get the final materials.

28.2.2 Characterization

The P-XRD analysis was performed by Rigaku X-ray diffractometer operating at 30 kV and 15 mA with a high-intensity beam of Cu-K_α (wavelength, $\lambda = 1.541$ Å) in the 2θ range of 5 to 70° and a scanning rate of 2° per minute. FESEM images were recorded by Carl Zeiss, Ultra 55 FESEM (Germany) operating at 20 kV. To perform CV, GCD and EIS measurements, Multi Autolab/M204 electrochemical workstation (Metrohm Autolab B.V. Netherlands) was used under ambient conditions in the three-electrode assembly.

28.2.3 Electrode Fabrication for Electrochemical Analysis

For electrode fabrication, first two (1 cm × 2 cm) pieces of graphite foil were cleaned with absolute ethanol. Then, slurries were prepared by mixing respective active materials with carbon black and polyvinylidene fluoride (PVDF) in (80:10:10) wt.% ratio. A few drops of NMP were added to the above mixtures and magnetically stirred for 8 h to make homogeneous suspensions. The obtained slurries were drop casted on cleaned graphite foil (1 cm × 1 cm) and dried at 60 °C overnight for the final electrode fabrication. The mass of the deposited material was ~1 mg, as calculated by the weight difference method.

28.3 Results and Discussion

Figure 28.1a shows the indexed P-XRD data of NiMn-LDH@HMTA and NiMn-LDH@urea. The (*hkl*) planes of all the diffraction peaks match well with that of typical layered structured hydrotalcite materials, similar to α-$Ni(OH)_2$ (JCPDS # 38–0715). The absence of any impurity reflection in the P-XRD profile confirms the formation of pure phase for both materials. The Debye–Scherrer calculation from the peak broadening of P-XRD data gives a crystallite size of 10.2 nm for NiMn-LDH@HMTA as compared to 15.2 nm for NiMn-LDH@urea. The interlayer spacing for the former (7.14 Å) is slightly lower than the latter (7.2 Å), as calculated from the interlayer spacing of the (003) plane.

FESEM images of NiMn-LDH@HMTA and NiMn-LDH@urea are shown in Fig. 28.1b–d. Both materials have a flower-like microstructure which is formed by the aggregation of various microplates. In the case of NiMn-LDH@urea, these plates are aggregated densely, making a close flower-like structure (Fig. 28.1b, c), while an open flower-like microstructure is obtained for NiMn-LDH@HMTA (Fig. 28.1d), due to loose stacking of the microplates. The relatively open microstructure of NiMn-LDH@HMTA may support the enhanced electrolyte penetration resulting in better electrode–electrolyte interface formation and ion diffusion, which will lead to improved electrochemical performance as compared to NiMn-LDH@urea with a compact microstructure.

The cyclic voltammetry (CV) analyses of the studied samples are carried out at the scan rate range of 10 $mVsec^{-1}$ to 100 $mVsec^{-1}$ in the potential range of 0 to 0.5 V (Fig. 28.2a, b). The presence of clear anodic and cathodic peaks with large voltage separation (greater than 0.1 V) confirms the battery type charge storage mechanism of the studied samples [5]. To ensure the above, log i is plotted against log υ and a linear curve fitting is obtained (Fig. 28.2c). The expected slopes for a battery type charge storage mechanism should be close to 0.5, while it should be close to 1 for a capacitor type charge storage mechanism [5]. The obtained slope of 0.47 for NiMn-LDH@HMTA and 0.53 for NiMn-LDH@urea substantiates the battery type charge storage mechanism in the samples. The specific capacitance/capacity values of 1144

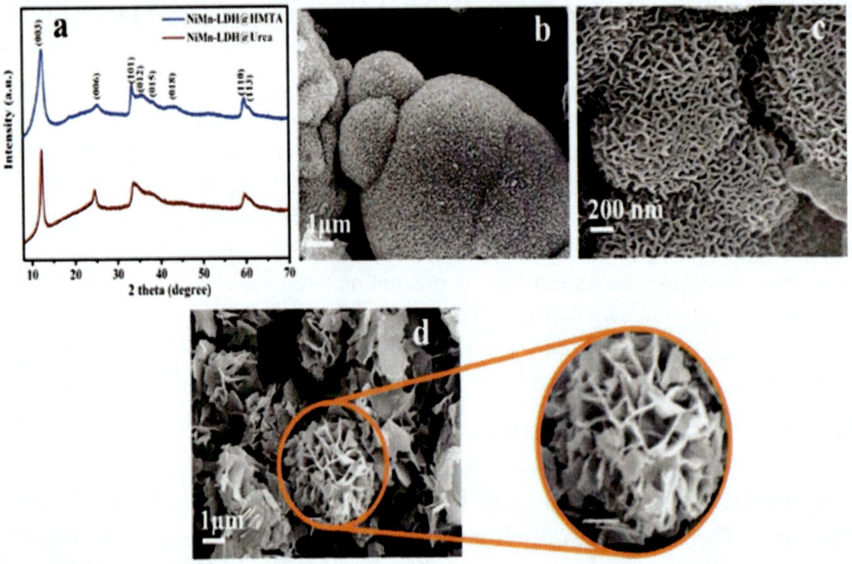

Fig. 28.1 **a** P-XRD patterns. FESEM images of (**b** & **c**) NiMn-LDH@urea and **d** NiMn-LDH@HMTA

(572), 917 (459), 761 (380), 589 (294), 483 (242), 377 (188) Fg^{-1} (Cg^{-1}) for NiMn-LDH@HMTA and 815 (408), 676 (338), 589 (294), 476 (238), 406 (203), 341 (171) Fg^{-1} (Cg^{-1}) for NiMn-LDH@urea are calculated at scan rates of 10, 20, 30, 50, 70 and 100 $mVsec^{-1}$, respectively (Fig. 28.2d) The high specific capacitances for NiMn-LDH@HMTA confirm its superiority over NiMn-LDH@urea, also consistent with its small crystallite size and open flower-like microstructure.

The GCD measurement of the samples is performed in the potential window of 0 to 0.4 V, where the current density ranges from 1 Ag^{-1} to 10 Ag^{-1} (Fig. 28.3a & 3b). The non-linear shape of the GCD curve again confirms the battery type charge storage mechanism consistent with the CV data. Among both materials, NiMn-LDH@HMTA shows a large discharge time (Fig. 28.3c), indicating its high specific capacitance, in agreement with the CV data. The specific capacitance/capacity of 1685 (657), 1302 (508), 1168 (455), 1023 (399), 905 (353), 787 (307) Fg^{-1} (Cg^{-1}) for NiMn-LDH@HMTA and 1218 (487), 882 (353), 698 (279), 532 (213), 416 (167), 294 (118) Fg^{-1} (Cg^{-1}) for NiMn-LDH@urea is obtained at 1, 2, 3, 5, 7 and 10 Ag^{-1}, respectively (Fig. 28.3d). The decrease in the specific capacitance with increasing current density is due to insufficient diffusion at high current densities. To check cyclic stability and its retention, 5000 GCD cycles (Fig. 28.3e) are performed at 10 Ag^{-1} for both materials. The capacity retention of 82% for NiMn-LDH@HMTA and 70% for NiMn-LDH@urea again confirms the superiority of NiMn-LDH@HMTA and its potential to be used as the cathode material for practical energy storage devices.

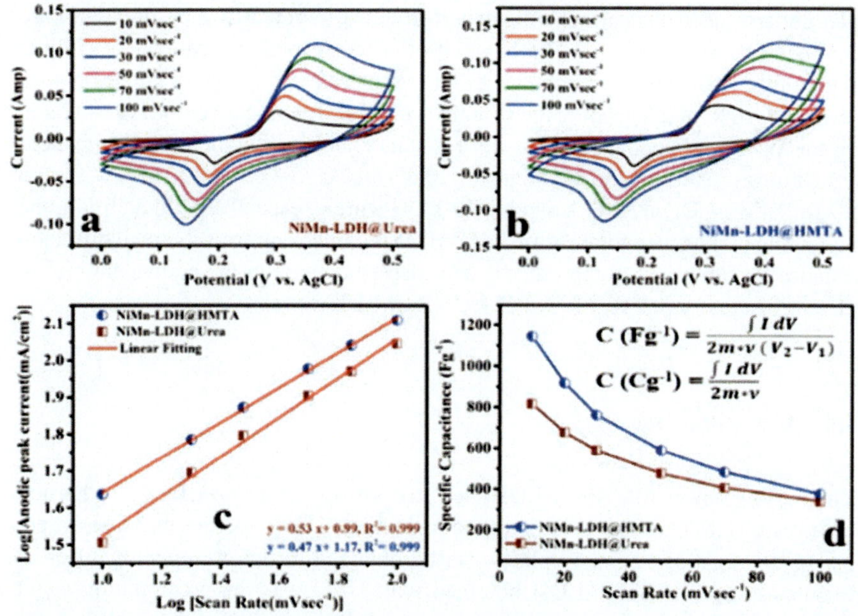

Fig. 28.2 CV curves for **a** NiMn-LDH@urea and **b** NiMn-LDH@HMTA. **c** The plot of Log υ versus Log i. **d** Specific capacitance dependence on scan rate

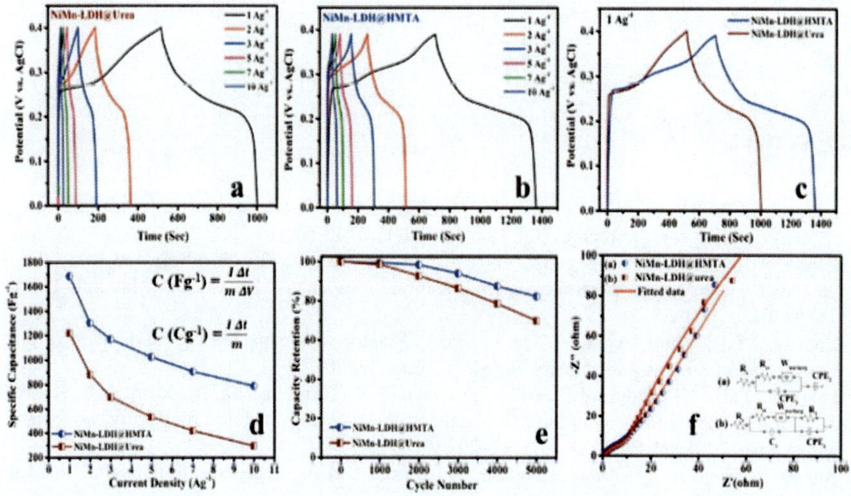

Fig. 28.3 GCD curves of **a** NiMn-LDH@urea and **b** NiMn-LDH@HMTA. **c** Comparison of GCD curves at 1 Ag^{-1}. **d** Specific capacitance versus current density data, **e** GCD cycle data and **f** Nyquist plots with corresponding equivalent circuit fitting

To understand the electrochemical performance better, EIS measurements are performed for both materials. The Nyquist plots, corresponding equivalent circuits and their fittings are shown in Fig. 28.3f. The semicircle in the high frequency region corresponds to R_s and R_{ct}, while the straight line in low frequency region is associated with the Warburg impedance, diffusion resistance and capacitance. The equivalent circuit fitting results in the R_s values of 0.47 and 0.75 Ω and R_{ct} of 7.5 Ω and 10 Ω for NiMn-LDH@HMTA and NiMn-LDH@urea, respectively. In addition, the linear fitting is less steep for NiMn-LDH@HMTA, which confirms its small diffusion resistance as compared to the other and is also evidenced in the equivalent circuit with an extra resistance R_1 of 11.4 Ω for NiMn-LDH@urea.

28.4 Conclusion

In conclusion, flower-like NiMn-LDHs are successfully synthesized by hydrothermal method using HMTA and urea as precipitating agents. This work reveals that different types of aggregations of microplates can be obtained in the presence of different precipitating agents. NiMn-LDH obtained with HMTA as the precipitating agent shows higher specific capacitance/capacity of 1685 Fg^{-1} (657 Cg^{-1}) at 1 Ag^{-1} due to its microstructure, smaller R_s (0.47 Ω), R_{ct} (7.5 Ω) and diffusion resistance. Realizing its high capacity and retention (82%) over 5000 GCD cycles, HMTA-mediated NiMn-LDH and its analogues could lead to hybrid energy storage devices for practical applications.

References

1. Lukatskaya, M.R., Dunn, B., Gogotsi, Y.: Multidimensional materials and device architectures for future hybrid energy storage. Nat. Commun. **7**(1), 1–13 (2016)
2. Brisse, A.L., Stevens, P., Toussaint, G., Crosnier, O., Brousse, T.: Ni $(OH)_2$ and NiO based composites: battery type electrode materials for hybrid supercapacitor devices. Materials **11**(7), 1178 (2018)
3. Zuo, W., Li, R., Zhou, C., Li, Y., Xia, J., Liu, J.: Battery-supercapacitor hybrid devices: recent progress and future prospects. Adv. Sci. **4**(7), 1600539 (2017)
4. Patel, R., Park, J.T., Patel, M., Dash, J.K., Gowd, E.B., Karpoormath, R., Mishra, A., Kwak, J., Kim, J.H.: Transition-metal-based layered double hydroxides tailored for energy conversion and storage. J. Mater. Chem. A **6**(1), 12–29 (2018)
5. Simon, P., Gogotsi, Y., Dunn, B.: Where do batteries end and supercapacitors begin? Science **343**(6176), 1210–1211 (2014)

Chapter 29
Investigation of Hot Corrosion Behaviour of Ytterbium Oxide Doped Gadolinium Zirconate Thermal Barrier Coating for Gas Turbines

M. Rajasekaramoorthy, A. Karthikeyan, A. Anderson, A. M. Kamalan Kirubaharan, and S. Anandh Jesuraj

Abstract The hot corrosion behaviour of Yb_2O_3 doped gadolinium zirconate ($Gd_2Zr_2O_7$, GZ) as a ceramic top coat for thermal barrier coatings (TBCs) applications is investigated in this study. The solid-state reaction approach was used to synthesize Yb_2O_3 doped GZ (9, 12, 15 weight percent Yb_2O_3). Prior to applying the ceramic top coat, the metallographically prepared Hastelloy X substrate was plasma sprayed with NiCrAlY metallic bond coat. Using the atmospheric plasma spray (APS) process, ceramic top coatings such as 9YbGZ, 12YbGZ, and 15YBGZ were deposited onto bond coated Hastelloy X substrates. To examine the hot corrosion behaviour of TBCs, molten salt mixture of Na_2SO_4 and V_2O_5 were applied to coated samples and subjected to high temperature at 1273 K for 50 h in a muffle furnace. The structural properties were investigated using the X-ray diffraction (XRD) technique. After the hot corrosion test, a high resolution—scanning electron microscopy (SEM) was utilised to evaluate the surface morphology and microstructural properties of the coatings. The hot corrosion resistance of the coatings and their performance were thoroughly reviewed.

Keywords Ytterbium oxide doped GZ · TBC · GZ · Hot corrosion · APS process

---

M. Rajasekaramoorthy · A. Karthikeyan (✉) · A. Anderson
School of Mechanical Engineering, Sathyabama Institute of Science and Technology, Chennai, India
e-mail: naveenkarthik2004@gmail.com

M. Rajasekaramoorthy
e-mail: rajasekaramoorthy23@gmail.com

M. Rajasekaramoorthy · A. M. K. Kirubaharan · S. A. Jesuraj
Centre for Nanoscience and Nanotechnology, Sathyabama Institute of Science and Technology, Chennai, India

A. M. K. Kirubaharan
Coating Department, Fun Glass - Centre for Functional and Surface Functionalised Glass, Alexander Dubcek University of Trencin, 91150 Trencin, Slovakia

A. Tewari et al. (eds.), *Proceedings of the 3rd International Conference on Advances in Materials Processing: Challenges and Opportunities*, Springer Proceedings in Physics 293, https://doi.org/10.1007/978-981-99-1971-0_29

29.1 Introduction

In order to enhance lifetime, thermal barrier coatings (TBCs) are applied to the gas turbine engine components which are exposed to extreme hot gaseous and corrosive environments [1]. Degradation of TBCs during service in gas turbine engines can be costly to repair and in extreme cases could be catastrophic. 7–8 wt. % of yttria stabilized zirconia (YSZ) is used as ceramic top coat material in TBCs for its excellent fracture toughness and cyclic life. However, YSZ suffers from undesirable phase transformation, sintering, hot corrosion, CMAS infiltration, etc. when operating at or above 1473 K for a prolonged duration [2]. Therefore, new ceramic top coat materials with desirable properties are highly required for advanced TBCs. Pyrochlore type of rare earth gadolinium zirconate, $Gd_2Zr_2O_7$ (GZ) has been identified as an alternative ceramic top coat material to YSZ because of their potential properties such as lower thermal conductivity (~2.1 Wm^{-1} K^{-1}) than that of YSZ (~1.1–1.7 Wm^{-1} K^{-1}), close thermal expansion co-efficient, better phase stability etc. [3–5].

However, GZ suffers from poor mechanical properties, which lead to reduced thermal cyclic performance when operated at 1473 K [6, 7]. Thus, current research is focused on finding a suitable dopant material to improve the mechanical properties of GZ. Few literatures [8–10] reported that, rare earth oxide materials such as Dy_2O_3, Er_2O_3 and Yb_2O_3 doped GZ provide improved mechanical properties than that of pure GZ. Lei et al. [11] have reported that Yb_2O_3 doped GZ showed lower thermal conductivity, higher thermal expansion co-efficient and improved cyclic performance than that of GZ.

Few techniques are available to develop the ceramic top coat. However, APS technique has been chosen in the current study to develop Yb_2O_3 doped GZ due to cost effectiveness and high deposition rate, which provides high porosities and lower thermal conductivity [12].

The hot corrosion behaviour study of thermal barrier coating is very important for understanding the life cycle of the TBCs at extreme conditions. Several literatures on hot corrosion study have been carried out by applying Na_2SO_4 and V_2O_5 paste and heat treating at high temperatures of about 850 °C, 900 °C etc. [13–15]. Thus, the present work is focused to develop the Yb_2O_3 doped GZ with various concentrations to investigate the effect of ytterbium oxide concentrations on the hot corrosive behaviour of the coatings. Thus, 9 wt. % of Yb_2O_3 doped GZ, 12 wt. % of Yb_2O_3 doped GZ and 15 wt. % of Yb_2O_3 doped GZ coatings were developed by using atmospheric plasma spray technique, and the developed coatings were subjected to hot corrosive test at 1273 K for 50 h in order to study the hot corrosive resistant behaviour of the coatings.

This research is aimed to improve the properties of gadolinium zirconate, which is one of the most promising ceramic materials to replace conventional yttria stabilized zirconia. With the improved fracture toughness and cyclic life of the doped GZ TBC, the gas turbines can operate at a higher temperature, which significantly improves the fuel efficiency and life cycle. This eventually reduces cost, which is very important in any economic situation.

29.2 Experiment

Hastelloy X substrates with a 20 mm diameter were metallurgically prepared and sand blasted with alumina particles ranging in size from 940 to 1080 μm. A metco 3 MB spray gun was used to plasma spray the NiCrAlY metallic bond layer. The plasma spray gun was used with argon primary gas and hydrogen secondary gas at flow rates of 65–70 LPM and 80–90 LPM, respectively at 500A current and 65–70 V voltage. The operational parameters were optimised and chosen from our prior study [16]. The Yb_2O_3 doped GZ were synthesised using solid state reaction. 9 wt. % Yb_2O_3 doped GZ (9YbGZ), 12 wt. % Yb_2O_3 doped GZ (12YbGZ) and 15 wt. % Yb_2O_3 doped GZ (15YbGZ) were synthesised and deposited using APS over the bond coat to a thickness 150–200 μm.

The coating's X-ray diffraction (XRD) pattern was recorded using a Bruker D8 Advanced (USA) equipment. Thermosceintific Apreo S high resolution scanning electron microscope (SEM) was used to examine high magnification images of the surface morphology of the coatings.

Molten salt hot corrosion test was carried out by applying a 4 mg/cm^2 paste of 55:45 ratio of Na_2SO_4: V_2O_5 salt on the surface of the TBC coated disc substrate along the centre of the disc at least 5 mm away from the edges. The salt applied samples were heat treated using a muffle furnace at 1000 °C isothermally for 50 h. The coatings were inspected every 5 h for failure by opening the furnace door and closed without allowing the temperature drop. Following the hot corrosion test, XRD and SEM analyses were done to investigate the coating's failure.

29.3 Result and Discussion

Figure 29.1 shows the air plasma sprayed YbGZ thermal barrier coatings deposited on Hastelloy X substrates and damaged coatings after hot corrosion test. As coated XRD pattern and XRD pattern of hot corrosion tested samples are plotted in Fig. 29.2.

The peaks on the as coated XRD are indexed and correspond with the standard JCPDS document for cubic GZ and cubic Yb_2O_3. Similarly, HR-SEM high magnification images of as coated and hot corrosion tested samples are shown in Fig. 29.3. 12YbGZ and 15YbGZ are more porous compared to 9YbGZ which are beneficial for improving thermal cycling performance during thermal cycling, better thermal insulation and also, is an important factor that helps to reduce thermal fatigue failure. The pores, cracks and splats are identified and marked on the HR-SEM micrograph in Fig. 29.3c, f.

The TBCs were subjected to 1000 °C with a corrosive mixture applied on their surface. The corrosive mixture at high temperature melts and reacts with the TBC. The 9YbGZ showed over 20% failure after 20 h followed by catastrophic failure after 25 h where the coating was completely peeled off and shattered. 12YbGZ suffered over 20% coating failure after 40 h. 15YbGZ suffered less than 20% failure after

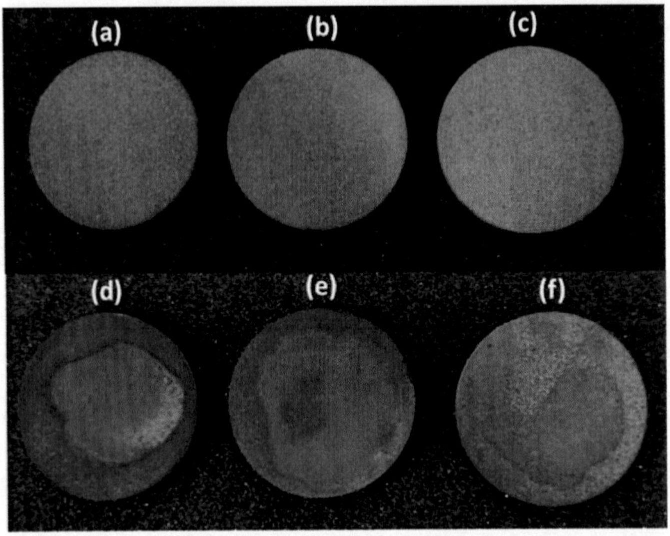

Fig. 29.1 Images of TBC before and after hot corrosion test at 1273 K **a** as coated 9YbGZ, **b** as coated 12YbGZ **c** as coated 15 YbGZ TBC **d** hot corrosion tested 9YbGZ **f** hot corrosion tested 12YbGZ and **f** hot corrosion tested 15 YbGZ

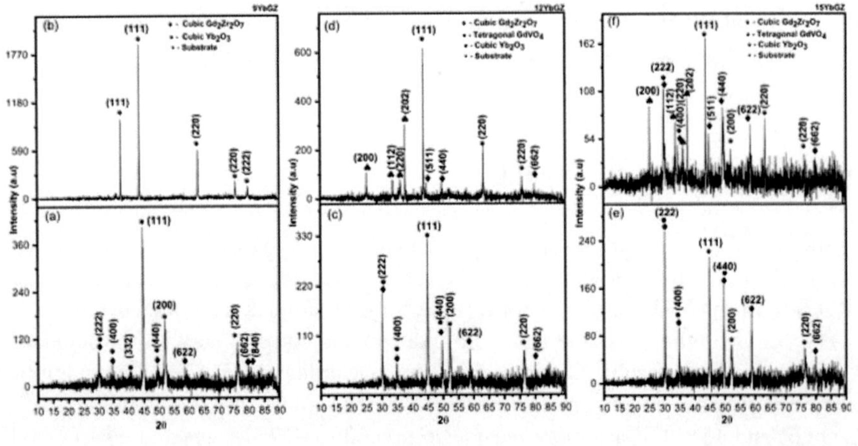

Fig. 29.2 XRD pattern of **a** as coated 9YbGZ **b** 9YbGZ after hot corrosion test **c** as coated 12YbGZ and **d** 12YbGZ after hot corrosion test **e** as coated 15YbGZ and **f** 15YbGZ after hot corrosion test

50 h. In the case of 15YbGZ and 12YbGZ, molten salt reacts with the GZ in the TBC and forms a new phase tetragonal $GdVO_4$, which is identified from the XRD patterns after hot corrosion test while only substrate peaks were obtained for 9YbGZ as most of the top coat reacted and eroded.

Also, the $GdVO_4$ crystals formed on the surface of the 15YbGZ and 12YbGZ can be identified from the HR-SEM micrographs. The HR-SEM micrograph clearly show

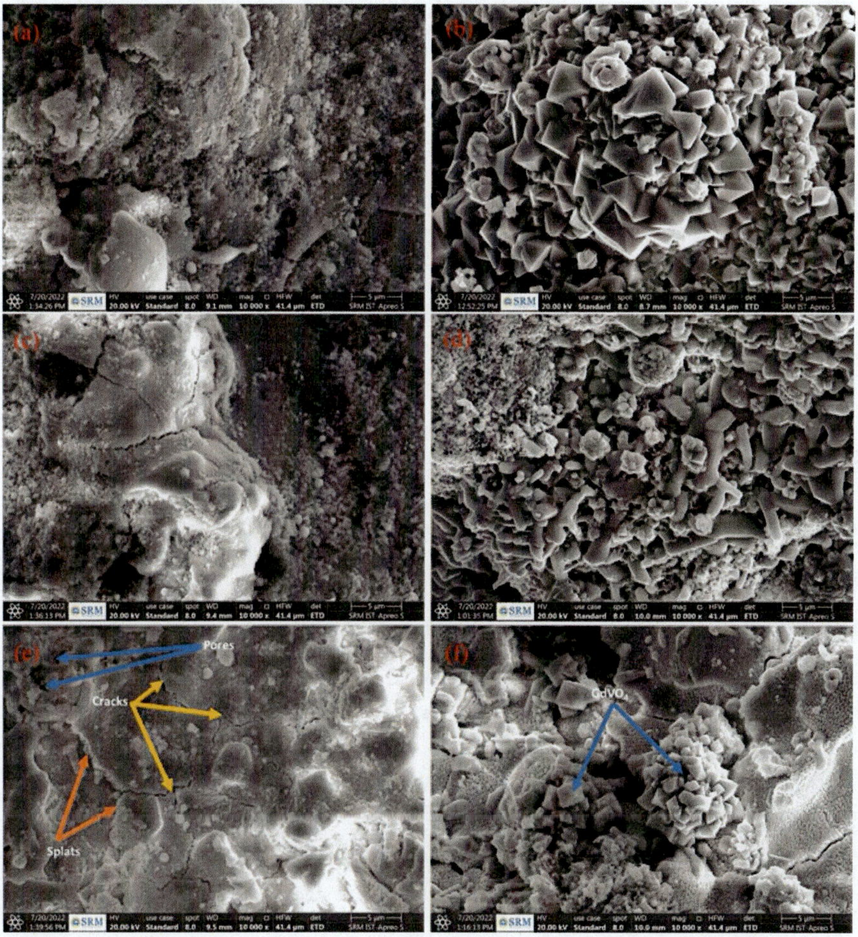

Fig. 29.3 High Magnification HR-SEM Micrograph of **a** as coated 9YbGZ **b** 9YbGZ after hot corrosion test **c** as coated 12YbGZ **d** 12YbGZ after hot corrosion test **e** as coated 15YbGZ and **f** 15YbGZ after hot corrosion test

the less reaction and lesser reaction product formed on the surface of the 15YbGZ compared to 12YbGZ. An increased amount of Yb_2O_3 in GZ improves the overall microstructure of the TBC and hence improved corrosion resistance.

29.4 Conclusion

Ytterbium oxide doped gadolinum zirconate thermal barrier coatings were deposited successfully on Hastelloy X substrates using air plasma spray coating technique. Molten salt hot corrosion test was performed on the TBCs. TBCs were characterized before and after hot corrosion test using XRD and HR-SEM. 15YbGZ TBC observed less than 20% coating failure even after 50 h of hot corrosion test. An increase in the doping concentration of Yb_2O_3 in GZ clearly enhances the TBC performance. For optimal performance, it is planned to increase the dopant concentration up to 20 % in future research.

Acknowledgements We authors acknowledge the Chancellor, President and Vice Chancellor, Sathyabama Institute of Science and Technology, Chennai, for providing the necessary facilities to carry out the research work. We acknowledge the XRD FACILITY at SRMIST set up with support from MNRE (Project No. 31/03/2014-15/PVSE-R&D), Government of India. We also acknowledge SRMIST for high resolution scanning electron microscope (HR-SEM) facility.

References

1. Saini, A.K., Das, D., Pathak, M.K.: Thermal barrier coatings -applications, stability and longevity aspects. Procedia Eng. **38**, 3173–3179 (2012)
2. Doleker, K.M., Karaoglanli, A.C.: Comparison of oxidation behavior of YSZ and Gd2Zr2O7 thermal barrier coatings (TBCs). Surf. Coat. Technol. **318**, 198–207 (2017)
3. Lashmi, P.G., Majithia, S., Shwetha, V., Balaji, N., Aruna, S.T.: Improved hot corrosion resistance of plasma sprayed YSZ/ Gd2Zr2O7 thermal barrier coating over single layer YSZ. Mater. Charact. **147**, 199–206 (2019)
4. Mahade, S., Li, R., Curry, N., Björklund, S., Markocsan, N., Nylén, P.: Isothermal oxidation behavior of Gd2Zr2O7/YSZ multilayered thermal barrier coatings. Int. J. Appl. Ceram. Technol. **13**(3), 443–450 (2016)
5. Mahade, S., Curry, N., Björklund, S., Markocsan, N., Nylén, P.: Thermal conductivity and thermal cyclic fatigue of multilayered Gd2Zr2O7/YSZ thermal barrier coatings processed by suspension plasma spray. Surf. Coat. Technol. **283**, 329–336 (2015)
6. Kumar, R., Cietek, D., Jiang, C., Roth, J., Gell, M., Jordan, E.H.: Influence of microstructure on the durability of gadolinium zirconate thermal barrier coatings using APS & SPPS processes. Surf. Coat. Technol. **337**, 117–125 (2018)
7. Guven Gok, M., Goller, G.: State of the art of gadolinium zirconate based thermal barrier coatings: design, processing and characterization. In: Methods for Film Synthesis and Coating Procedures, IntechOpen (2022)
8. Shen, Z., Liu, G., Liu, Z., He, L., Mu, R., Wang, L.: Dy doped Gd2Zr2O7 thermal barrier coatings: thermal expansion coefficient, microstructure and failure mechanism. Appl. Surface Sci. Adv. **6**, 100174 (2021)
9. Guo, L., Zhang, C., Li, M., Sun, W., Zhang, Z., Ye, F.: Hot corrosion evaluation of Gd2O3-Yb2O3 co-doped Y2O3 stabilized ZrO2 thermal barrier oxides exposed to Na2SO4+V2O5 molten salt. Ceram. Int. **43**(2), 2780–2785 (2017)
10. Dong, S., Zhang, F., Li, N., Zeng, J., Liang, P., Zhang, H., Cao, X.: Thermal radiation and cycling properties of (Ca, Fe) or (Sr, Mn) co-doped La2Ce2O7 coatings. J. Eur. Ceram. Soc. **50**(5), 2020–2029 (2020)
11. Guo, L., Guo, H., Peng, H., Gong, S.:Thermophysical properties of Yb2O3 doped Gd2Zr2O7 and thermal cycling durability of (Gd0.9Yb0.1)2Zr2O7/YSZ thermal barrier coatings. J. Europ. Ceramic Soc. **34**(5), 1255–1263 (2014)
12. Guo, L., Li, M., He, S., Zhang, C., Wang, Q., Ye, F.: Preparation and hot corrosion behavior of plasma sprayed nanostructured Gd2Zr2O7 -LaPO4 thermal barrier coatings. J. Alloy. Compd. **698**, 13–19 (2017)

13. Afrasiabi, A., Saremi, M., Kobayashi, A.: A comparative study on hot corrosion resistance of three types of thermal barrier coatings: YSZ, YSZ+Al2O3 and YSZ/Al2O3. Mater. Sci. Eng., A **478**(1–2), 264–269 (2008)
14. Ahmadi, M.S., Shoja-Razavi, R., Valefi, Z., Jamali, H.: Evaluation of hot corrosion behavior of plasma sprayed and laser glazed YSZ–Al2O3 thermal barrier composite. Opt. Laser Technol. **111**, 687–695 (2018)
15. Ozgurluk, Y., Doleker, K., M. and Karaoglanli A. C.,: Hot corrosion behavior of YSZ, Gd2Zr2O7 and YSZ/ Gd2Zr2O7 thermal barrier coatings exposed to molten sulfate and vanadate salt. Appl. Surf. Sci. **438**, 96–113 (2018)
16. Kamalan Kirubakaran, A.M., Anderson, A., Rajasekaramoorthy, M., Rama Narasimham, P., Subramanyam, P.:Hot corrosion behaviour of plasma sprayed YSZ and Gadolinium oxide doped YSZ thermal barrier coatings. Mater. Today Proc., 6–11 (2020)

Chapter 30
Thermal and Flame Retardancy Behavior of Eggshell Filler Reinforced Polypropylene Composites

Sandeep Gairola, Shishir Sinha, and Inderdeep Singh

Abstract Eggshell waste is the most abundant natural waste in nature and is created in massive amounts due to the everyday consumption of eggs. Most of this material is being discarded, even though it has multidisciplinary applications. The current research focused on processing and investigation of green composites' thermal and flame retardancy behavior by utilizing the eggshell (ES) filler with ammonium polyphosphate (APP) (an intumescent flame retardant) in polypropylene. The optimization of ES filler and APP content in the developed composites was examined by varying the overall loading of 10, 20, and 30 wt. %, keeping an ES to APP ratio constant as 1:1. The thermal stability and flame retardancy were investigated with thermogravimetric analysis, Ul-94 horizontal burning (HB), and limiting oxygen index (LOI). The developed composites have shown increased thermal stability and flame retardancy. The HB rate was reduced, and LOI increased by 37.38% and 39.87%, respectively, as compared to PP for 20 wt.% loading. The possibility of utilizing the ES filler as a flame retardant will further broaden the application potential of this material.

Keywords Eggshell · Biofiller · Flame retardant · APP · Polypropylene

S. Gairola (✉) · S. Sinha · I. Singh
Centre of Excellence in Disaster Mitigation and Management, Indian Institute of Technology Roorkee, Roorkee, Uttarakhand 247667, India
e-mail: sandeepgairola.12345@gmail.com; s_gairola@dm.iitr.ac.in

S. Sinha
e-mail: shishir@ch.iitr.ac.in

I. Singh
e-mail: inderdeep.singh@me.iitr.ac.in

S. Sinha
Department of Chemical Engineering, Indian Institute of Technology Roorkee, Roorkee, Uttarakhand 247667, India

I. Singh
Department of Mechanical and Industrial Engineering, Indian Institute of Technology Roorkee, Roorkee, Uttarakhand 247667, India

A. Tewari et al. (eds.), *Proceedings of the 3rd International Conference on Advances in Materials Processing: Challenges and Opportunities*, Springer Proceedings in Physics 293, https://doi.org/10.1007/978-981-99-1971-0_30

30.1 Introduction

Eggshell (ES) is a natural bio-filler containing 95 percent calcium carbonate ($CaCO_3$) and only 5% organic content. The food processing sector produces around 250,000 tons of ES yearly [1]. The utilization of eggshells as a bio filler in developing polymer composites has resulted in better thermal properties than mechanical properties. Therefore, ES can be used as a flame-retardant (FR) material to develop polymer composites. Polypropylene (PP) is a versatile thermoplastic commonly used in automobile components, household appliances, and other industrial applications. The manufacturing of PP composites has attracted considerable interest due to their distinctive characteristics, such as their excellent mechanical capabilities, thermal stability, and lower overall product cost. However, the low thermal stability of PP limits its applications, and therefore to fulfill stringent technical and structural requirements, PP is seldom utilized in its pure form and is frequently turned into a composite material by the addition of fillers or reinforcements [2, 3]. $CaCO_3$ is one of the most widely used mineral fillers since it is inexpensive and easily available in usable form. ES has recently emerged as a viable filler for PE/PP composites used in various applications. Krishnan et al. [1] investigated the thermal and mechanical behavior of PP/ES composites. Their study reported that the composites exhibit high thermal degradation temperatures while lower yield strength with improved young modulus compared to neat PP. Ahmed et al. [2] explored the thermo-mechanical behavior of $CaCO_3$ (derived from eggshell) filled PP composites. It was reported that adding 30-wt % ES filler to PP delayed the flame propagation time by 99 s. Raj et al. [4] studied the thermomechanical behavior of PP/ES composites. The study showed that PP composites with modified ES had better mechanical properties than unmodified ES and $CaCO_3$ composites. Zhisheng et al. [5] investigated the influence of ES on the thermal stability, flame-retardant, and smoke-suppression properties of epoxy composites. It was found that the incorporation of ES improves the LOI value and thermal stability of the epoxy. Maysa et al. [6] investigated the influence of ES filler on thermal and flammability behavior of High-Density Polyethylene (HDPE). It was reported that the thermal stability and flame retardancy of composites increased by adding ES to HDPE. However, the ES cannot perform well alone in the polymeric system. Therefore, incorporating ES as a bio-filler with other intumescent fire-retardant materials has been reported to provide excellent synergistic performance in anti-oxidation properties, fire protection, thermal stability, and leads to the formation of a compact and uniform intumescent char layer. The use of the intumescent flame retardant in polymeric materials is one of the efficient ways to develop flame-resistant polymeric composite materials. Urtekin et al. [7] investigated the flammability behavior of PLA by incorporating ES and intumescent flame retardants (IFR), such as ammonium polyphosphate (APP) and pentaerythritol (PER). It was reported that 10 wt.% ES filler improves the LOI to 34.1 with a V0 rating. Younis et al. [3] investigated the effect of the addition of mono-ammonium phosphate (MAP) and magnesium hydroxide (MOH) on the properties of recycled waste polypropylene (WPP) and waste chicken eggshell (WCES) composites. The authors reported that incorporating ES improved the LOI value to 20.8 from 17.9. Asanka et al. [8] investigated the effect of ES and lignin with APP in the flammability behavior of epoxy composites and found improvement in the flammability properties.

From the reported work, there is still a wide scope for utilizing the ES biofiller with APP in polypropylene composites for developing an in-depth understanding for commercial applications. Therefore, it is worthwhile to explore the effect of APP with ES on the thermal and flame retardancy behavior of PP/ES composites.

30.2 Materials and Methods

30.2.1 Sample Preparation

The composite samples were prepared as shown in Fig. 30.1. The ES was thoroughly washed and cleaned several times with hot water to ensure the removal of impurities present, followed by drying at 80 °C for 24 h. Thereafter, ES and APP were mixed with PP and melt blended at 45 rpm in a single screw extruder with temperatures in the different heating zones as 175 °C, 175 °C, and 170 °C, from the hopper to die section. The extruded strands were passed through a water bath at 30 °C, followed by pelletizing into small pellets using a pelletizer. The developed pellets were then used to develop the final samples using an injection molding machine (Make: Electronica Plastic Machine Ltd. ENDURA 60) with different loadings, as shown in Fig. 30.1.

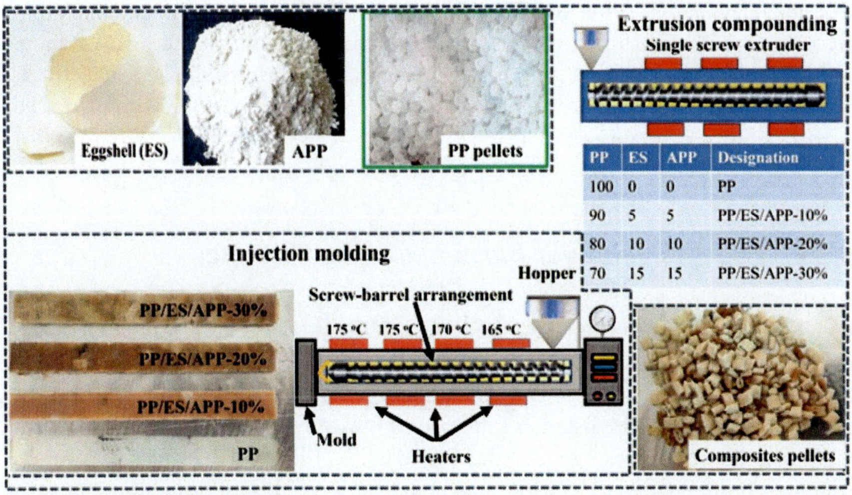

Fig. 30.1 Processing route of composite samples

30.2.2 Characterization Techniques

The developed samples were characterized using thermogravimetric analysis (TGA) (Make: 6300 EXSTAR), Ul-94 horizontal cum vertical flame tester (Make: Fugen Automation), and Limiting Oxygen Index (LOI) Tester (Make: Fugen Automation). TGA was performed at 10 °C/min in a nitrogen atmosphere from 30 °C to 700 °C.

30.3 Results and Discussion

30.3.1 Thermogravimetric Analysis

The changes in the thermal stability of the developed composites incorporated with ES and APP are shown in Fig. 30.2 and Table 30.1. Figure 30.2a, b show the thermal weight decomposition curves with respect to time. Under a nitrogen atmosphere, the ES filler mainly decomposes in three stages. The first stage has a low mass loss assigned to the decomposition of organic components in the ES, the second stage decomposes a part of organic components in the ES, and the third stage shows major decomposition, which is primarily attributed to the thermal decomposition of $CaCO_3$ concomitant with the formation of carbon dioxide and calcium oxide. Composites loaded with 20 wt.% ES/APP showed the lowest weight decomposition and highest char residue of about 22.8%. It could be due to the formation of thermally stable calcium metaphosphate and calcium phosphate by interaction between APP and $CaCO_3$, which serves as an efficient physical barrier against the mass transfer, thus effectively increasing the residual weight [5]. The major weight loss of about 75% took place for 20 wt.% ES/APP in 45 min. The results showed that developed composites with ES/APP recorded lower weight decomposition with increased thermal stability as compared to neat PP.

30.3.2 UL-94 Horizontal Burning and LOI Testing

The UL-94 and LOI test results for the PP/ES/APP samples are shown in Fig. 30.2c. The developed composites achieved reduced dripping, reduced burning rate, and increased LOI for all samples. The average horizontal burning time and the burning rate were recorded and are listed in Table 30.1. However, the maximum improvement in flammability properties was recorded for 20 wt.% of ES/APP. The neat PP showed the highest burning rate of 26.16 mm/min and the 20 wt.% ES/APP/PP bio-composite recorded a lowest value of 16.38 mm/min. This could be attributed to the better dispersion of the ES/APP particles in the PP matrix. The ES fillers act as a heat sink while APP act as an intumescent agent; therefore, a synergistic effect was recorded after incorporating both the fillers.

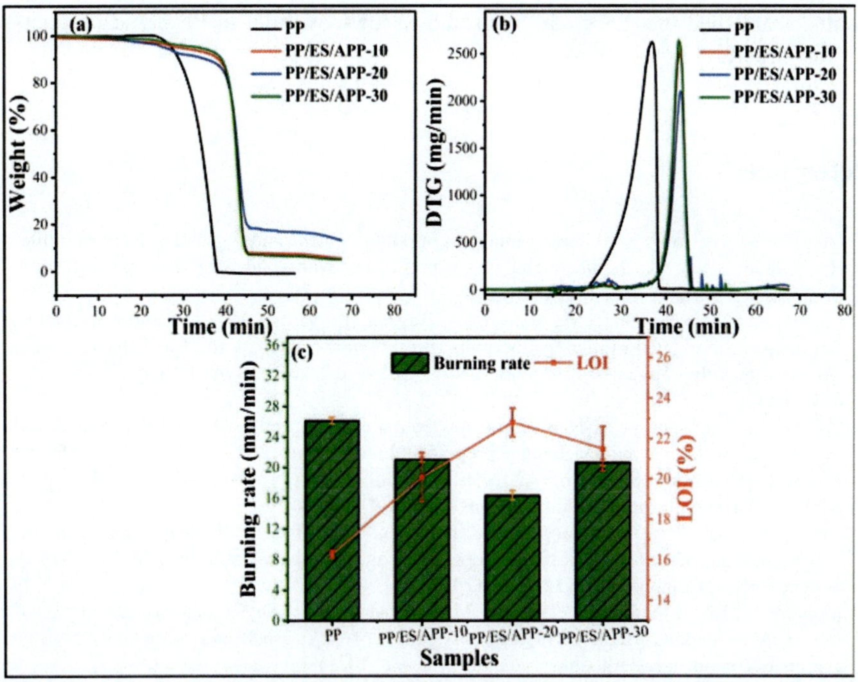

Fig. 30.2 **a** Thermal weight decomposition, **b** DTG curves, and **c** flammability behavior of ES/APP composites

Table 30.1 TGA data for developed composites

Samples	t_{10}	t_{50}	t_{75}	Char residue	DTG_{max}	Burning rate	LOI
	(min)			(%)	(mg/min)	(mm/min)	(%)
PP	29	34	36	0	2.618	26.16	16.3
PP/ES/APP-10	39	42	44	14.2	2.517	21.08	20.1
PP/ES/APP-20	34	43	45	14.5	2.641	16.38	22.8
PP/ES/APP-30	38	42	43	20.1	2.091	20.68	21.5

30.4 Conclusions

Waste ES was utilized with APP for the preparation of flame-retardant PP composites. The flame retardancy of the prepared composites was improved with the addition of ES/APP filler. The highest thermal stability (char residue 22.8%) and flame retardancy (HB: 16.38 mm/min and LOI: 22.8) were recorded with 20 wt.% of filler content as compared to neat PP. The results showed that the filler positively obstructed the flame propagation through the composite, the burning rate was reduced by 37.38% and LOI increased by 39.87% as compared to neat PP samples. The results established that ES could be used as a low-cost filler in PP-based products for non-structural applications.

References

1. Iyer, K.A., Torkelson, J.M.: Green composites of polypropylene and eggshell: effective biofiller size reduction and dispersion by single-step processing with solid-state shear pulverization. Compos. Sci. Technol. **102**, 152–160 (2014)
2. Hassen, A.A., Dizbay-Onat, M., Bansal, D., Bayush, T., Vaidya, U.: Utilization of chicken Eggshell Waste as a Bio-Filler for thermoplastic Polymers: thermal and Mechanical characterization of Polypropylene Filled With naturally derived caco3. Polym. Polym. Compos. **23**, 653–662 (2015)
3. Younis, A.A., El-Wakil, A.A.: New composites from waste polypropylene/eggshell characterized by high flame retardant and mechanical properties. Fibers Polym. 2021 2212 2021 **22**, 3456–68
4. Kumar, R., Dhaliwal, J.S., Kapur, G.S.: Shashikant. Mechanical properties of modified biofiller-polypropylene composites. Polym. Compos.**35**, 708–14 (2014)
5. Xu, Z., Chu, Z., Yan, L., Chen, H., Jia, H., Tang, W.: Effect of chicken eggshell on the flame-retardant and smoke suppression properties of an epoxy-based traditional APP-PER-MEL system. Polym. Compos. **40**, 2712–2723 (2019)
6. Mohamed, M.A., Mounir, R., EL-Zayat, M.M., Raslan, H.A.: Physico-mechanical comparative study on gamma irradiated high density polyethylene/eggshell and commercial calcium carbonate composites. Radiochim Acta **109**, 485–92 (2021)
7. Urtekin, G., Hazer, S., Aytac, A.: Effect of eggshell and intumescent flame retardant on the thermal and mechanical properties of plasticised PLA. Plast Rubber Compos. **50**, 127–136 (2021)
8. Basnayake, A.P., Hidalgo, J.P., Villacorta, B., Dunn, K.G., Heitzmann, M.T.: Investigation of ammonium polyphosphate dilution with ground eggshells and lignin through the study of natural fibre composite flammability. Fire Mater (2022)

Chapter 31
Photovoltaic Module Recycling Processes: A Review

Harish Trivedi, Arunabh Meshram, and Rajeev Gupta

Abstract Photovoltaic module (PVM) is a very popular energy device for green and renewable energy. Recycling of end of life (EoL) PVMs is becoming a necessity due to excessive e-waste generation. PVMs contain many precious and hazardous metals. Manufacturing of materials used in PVMs is highly energy intensive so effective recycling will increase the sustainability of the PVMs industry. There are many recycling processes available such as physical, chemical, thermal, and hybrid processes, however, optimization is required to handle huge e-waste which is expected by EoL PVMs in upcoming decades.

Keywords Photovoltaic modules · Recycling · Silicon (Si) wafer · Metal extraction · E-waste

31.1 Introduction

Renewable energy must be motivated to minimize the use of traditional energy sources [1]. The PVM is a very efficient, low-cost, and maintenance-free device. Observing the manufacturing of the PVMs, the global e-waste amount will rise to 60–78 million tonnes by 2050 [2, 3]. The huge e-waste pile-up due to EoL PVMs expected years ahead, is the chief motivation for recycling. The average life of dominantly used silicon cells is 20–30 years [4]. It is projected that recycling can save 60% energy of PVMs manufacturing and should become a reason for stakeholders' attraction and sustainability of PVMs industries. According to legal norms 80 wt.% recycling of e-waste is mandatory and it can be satisfied by recycling glass and aluminum [5]. There are many precious and hazardous metals being used to make

H. Trivedi (✉) · R. Gupta
Materials Science Programme, IIT Kanpur, Kanpur, India
e-mail: harishis@iitk.ac.in

A. Meshram
Department of Materials Science & Engineering, IIT Kanpur, Kanpur, India

A. Tewari et al. (eds.), *Proceedings of the 3rd International Conference on Advances in Materials Processing: Challenges and Opportunities*, Springer Proceedings in Physics 293, https://doi.org/10.1007/978-981-99-1971-0_31

solar cells and contain a total of 5 wt.% [5, 6]. The high demand for materials used in various wafer and thin-film solar cells can be partially fulfilled by recycling [5, 7].

31.2 Materials Used in Various PVMs

First-generation solar cells are wafer-based and made of crystalline silicon (c-Si) and GaAs. Silicon solar cells containing Si, Ag, Al, Cu, Pb, Sn, etc. are dominating the market [8]. Second-generation thin film solar cells contain Cd, Te, Cu, In, Se, Ga, Ge, As, etc. Third-generation solar cells are perovskite, quantum dot, dye-sensitized solar cells (DSSCs), organic polymer solar cells, and many more, containing elements such as Ti, Sn, Pb, Cs, Cu, Ag, etc. [9, 10]. In addition, Ethylene–vinyl acetate (EVA) is commonly used for encapsulation, and Tedlar® which is a fluorinated polymer used for backside support. Glass slab is used for environmental protection and an Al frame is used to frame the whole solar cell structure [11].

31.3 Recycling Processes

Due to the absence of an effective recycling process, the major part of EoL PVMs is landfilled or discarded [5, 12]. According to the waste management hierarchy, disposal is the last step to follow [12]. When recycled materials are used for a higher-value application, it is known as upcycling and for a lower-value application, it is known as the downcycling process [13]. As illustrated in Fig. 31.1 there are broadly three processes named chemical, thermal, and mechanical and their hybrid process such as thermo-chemical, thermo-mechanical, and chemo-mechanical are being used in the recycling of EoL PVMs. The main motive of recycling is the maximum recovery of materials and reduce effluent and toxic gas generation [14].

31.3.1 Physical Process

It comprises mechanical processes such as the separation of the Al frame and other accessories, and the comminution of EoL PVMs. Physical processes such as grinding to remove antireflection coating (ARC) and emitter layer of the silicon wafer, separation of the EVA layer by sonication, wire saw device, and "Hot knife" used for the upcycling process [12, 15]. To recycle CdTe and CIGS solar cells wet-mechanical and thermo-mechanical methods are used. Physical processes are simple, cost-effective, chemical-free, and industrial-scale however high emissions, energy, and contamination of recycled materials are the drawbacks [16].

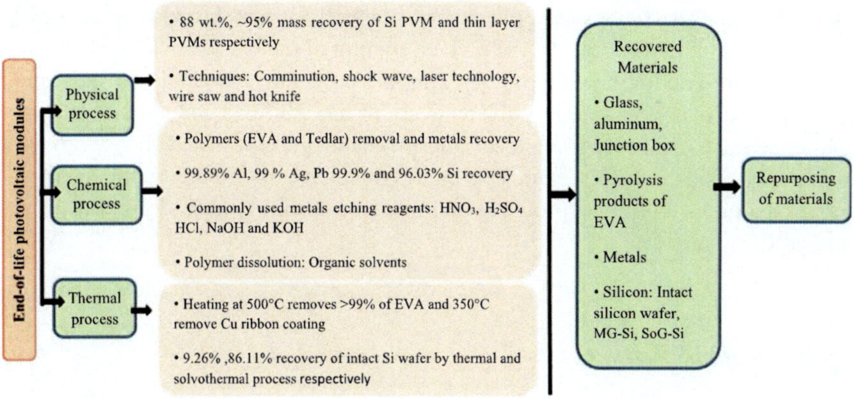

Fig. 31.1 Overview of end-of-life (EoL) photovoltaic modules (PVMs) recycling [17–21]

31.3.2 Chemical Process

Organic solvents are used in the dissolution of EVA. Commonly used organic solvents are tetrahydrofuran (THF), toluene, D-Limonene, and chloroform [22, 23]. The chemical reactions of organic solvents are different with EVA at the middle and joining surface to glass and solar cells. The middle part of EVA is non-crosslinked and gets dissolved whereas the joining surface of EVA is crosslinked and swells in the presence of an organic solvent [24]. Acids and bases such as HNO_3, HCl, HF, H_2SO_4, and KOH/NaOH are being used for the etching of metals [15]. The etching of metals is optimized by the concentration of reagent, etching time, and temperature [23, 25]. Several leaching reagents and combinations, a well-established leaching technique, and several organic solvents for EVA removal are available through chemical processes. Hazardous chemicals, silicon wafer thickness reduction, effluent generation, hazardous gas and fume emissions, and expertise and management are drawbacks.

31.3.3 Thermal Process

Pyrolysis of EVA occurs in two steps. In the first step, 30% weight loss occurs at 260–370 °C and acetic acid is the main product. In the second step, 90% weight loss occurs at 370–480 °C and the product are hydrocarbons such as ethane, propene, hexane, etc. [19]. Thermal treatment of EVA creates a huge amount of gases that can damage the silicon wafer. The ideal heating rate is 15 °C and the temperature is 480 °C to recover the maximum amount of intact silicon wafers [15]. Electrochemical heating of EVA also can be used for the separation of solar cell from glass and backsheet [26]. Thermal treatments are also used to separate Pb, Sn, and Ag from the surface of the Cu ribbons used to connect solar cells. When the vertically hanged Cu ribbon is heated up to 350 °C, the surface coating of the Cu ribbon is melted and collected due to gravitational force [27]. Thermal processes are viable for the rapid and efficient recovery of materials, the recovery of intact silicon wafers, the removal of polymer traces following chemical and physical processes, and the delamination of the backside polymer layer. This energy-intensive, high-temperature process degrades materials thermally, produces enormous quantities of hazardous gases, and increases CO_2 emissions.

31.3.4 Hybrid Process

A combination of the physical, chemical and thermal processes is required to achieve a quick, cost-effective, and environmentally friendly process. EVA can be dissolved either in organic solvents or through the physical process such as sonication and microwave heating but still, some traces of EVA can remain at the surface of the silicon wafer. These traces of EVA can be removed by heat treatment so these processes of removing EVA are considered thermo-chemical, and thermo-physical processes respectively [25, 28]. To remove the surface layers of the silicon wafer mechanical grinding is used which is an example of a physio-chemical process [16]. In the down-cycling process, solar cells are first comminuted by physical treatment followed by thermal treatment to remove polymers. Finally, chemical treatment is done for the etching of metals [29]. The thermo-mechanical processes are applied for the recycling of CdTe and CIS solar cells [16]. These processes offer a methodical way to recycling. However, they include hazards such as chemical flammability, undesirable chemical waste and emission, and expertise to examine process safety.

31.4 Repurposing of Recycled Materials

Materials can be used directly or after some refurbishments. The reuse of materials depends upon their recovery state and chemical composition [30]. The reuses of some important metals and materials are illustrated in Fig. 31.2. Commercialization of recycled PVM materials will support recycling economically and reduce the energy payback time (EPBT) of PVMs [12].

31.5 Conclusion

To efficiently recover valuable and hazardous elements from EoL PVMs, the recycling process must be an optimal combination of physical, thermal, and chemical processes. PVMs recyclers require sophisticated technology to recover upcycling

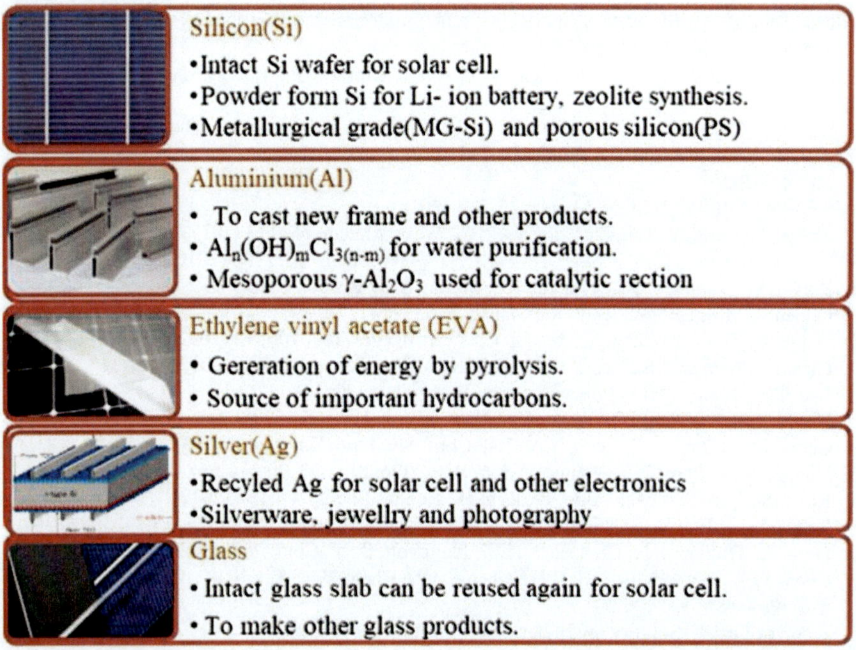

Fig. 31.2 Repurposing of EoL PVMs recycled materials for efficient reutilization [12, 15, 31, 32]

products and repurpose recycled materials in PVMs and other fields. Reclaimed materials can satisfy the raw material demand for PVMs manufacturing in the near future. Recycling will be more sustainable if energy consumption, effluent production, and gas emissions are reduced, while the amount of e-waste treatment, recovery rate, and income are maximized.

References

1. Castillo Esparcia, A., López Gómez, S.: Sustain. **13** (2021)
2. Granata, G., Pagnanelli, F., Moscardini, E., Havlik, T., Toro, L.: Sol. Energy Mater. Sol. Cells **123**, 239 (2014)
3. Walzberg, J., Carpenter, A., Heath, G.A.: Nat. Energy **6**, 913 (2021)
4. Doi, T., Tsuda, I., Unagida, H., Murata, A., Sakuta, K., Kurokawa, K.: Sol. Energy Mater. Sol. Cells **67**, 397 (2001)
5. Deng, R., Chang, N., Lunardi, M.M., Dias, P., Bilbao, J., Ji, J., Chong, C.M.: Prog. Photovoltaics Res. Appl. **29**, 760 (2021)
6. Mahmoudi, S., Huda, N., Alavi, Z., Islam, M.T., Behnia, M.: Resour. Conserv. Recycl. **146**, 1 (2019)
7. La Mantia, F.P.: Prog. Rubber, Plast. Recycl. Technol. **20**, 11 (2004)
8. Corcelli, F., Ripa, M., Leccisi, E., Cigolotti, V., Fiandra, V., Graditi, G., Sannino, L., Tammaro, M., Ulgiati, S.: Ecol. Indic. **94**, 37 (2018)
9. Sharma, S., Jain, K.K., Sharma, A.: Mater. Sci. Appl. **06**, 1145 (2015)

10. Ranabhat, K., Patrikeev, L., evna Revina, A.A., Andrianov, K., Lapshinsky, V., Sofronova, E.: J. Appl. Eng. Sci. **14**, 481 (2016)
11. Latunussa, C.E.L., Ardente, F., Blengini, G.A., Mancini, L.: Sol. Energy Mater. Sol. Cells **156**, 101 (2016)
12. Farrell, C.C., Osman, A.I., Doherty, R., Saad, M., Zhang, X., Murphy, A., Harrison, J., Vennard, A.S.M., Kumaravel, V., Al-Muhtaseb, A.H., Rooney, D.W.: Renew. Sustain. Energy Rev. **128**, 109911 (2020)
13. Wilson, M.: J. Prod. Brand Manag. **25**, 394 (2016)
14. Zhang, S., Ding, Y., Liu, B., Chi Chang, C.: Waste Manag. **65**, 113 (2017)
15. Park, J., Kim, W., Cho, N., Lee, H., Park, N.: Green Chem. **18**, 1706 (2016)
16. Berger, W., Simon, F.G., Weimann, K., Alsema, E.A.: Resour. Conserv. Recycl. **54**, 711 (2010)
17. Gönen, Ç., Kaplanoğlu, E.: Waste Manag. Res. **37**, 412 (2019)
18. Deng, R., Zhuo, Y., Shen, Y.: Resour. Conserv. Recycl. **187**, 106612 (2022)
19. Dias, P., Javimczik, S., Benevit, M., Veit, H.: Waste Manag. **60**, 716 (2017)
20. Lee, T.D., Ebong, A.U.: Renew. Sustain. Energy Rev. **70**, 1286 (2017)
21. Xu, X., Lai, D., Wang, G., Wang, Y.: Chem. Eng. J. **418**, 129457 (2021)
22. Chen, W.S., Chen, Y.J., Chen, Y.A.: IOP Conf. Ser. Earth Environ. Sci. **291** (2019)
23. Huang, W.H., Shin, W.J., Wang, L., Sun, W.C., Tao, M.: Sol. Energy **144**, 22 (2017)
24. Kang, S., Yoo, S., Lee, J., Boo, B., Ryu, H.: Renew. Energy **47**, 152 (2012)
25. Radziemska, E., Ostrowski, P., Cenian, A., Sawczak, M.: Ecol. Chem. Eng. S **17**, 385 (2010)
26. Doni, A., Dughiero, F.: Conf. Rec. IEEE Photovolt. Spec. Conf. 757 (2012)
27. Chen, W.S., Chen, Y.J., Lee, C.H., Cheng, Y.J., Chen, Y.A., Liu, F.W., Wang, Y.C., Chueh, Y.L.: Processes **9**, 1 (2021)
28. Kim, Y., Lee, J.: Sol. Energy Mater. Sol. Cells **98**, 317 (2012)
29. Dias, P.R., Benevit, M.G., Veit, H.M.: Waste Manag. Res. **34**, 235 (2016)
30. Redlinger, M., Eggert, R., Woodhouse, M.: Sol. Energy Mater. Sol. Cells **138**, 58 (2015)
31. Strachala, D., Hylský, J., Vaněk, J., Fafilek, G., Jandová, K.: Acta Montan. Slovaca **22**, 257 (2017)
32. Depeursinge, A., Racoceanu, D., Iavindrasana, J., Cohen, G., Platon, A., Poletti, P.-A., Muller, H.: Artif. Intell. Med. **229**, ARTMED1118 (2010)

Chapter 32
Optimization of Leaching Parameters for Recovery of Tantalum from Waste Tantalum Capacitors

Munmun Agrawal, Rohit Jha, Kamalesh K. Singh, and Randhir Singh

Abstract Recycling of tantalum seeking much attention in the recent past, considering its criticality and associated supply risk along with increased demand. Tantalum has become an element of choice in the modern electronic industry due to its superior capacitance in much compact size—a prime requirement of today's technological-driven society. This serves electronic waste as the best secondary resource for tantalum recovery. However, the recovery of tantalum from the capacitor is still developing and facing challenges associated with the separation of tightly covered mold resin containing silica and the dissolution of tantalum in fluoride-free media. In the present investigation, we have developed a hydrometallurgical method to recover tantalum from tantalum capacitors. The tantalum capacitor was first subjected to pre-treatment to separate the encapsulated silica-containing mold resin from the metallic-rich part. After this, the epoxy resin-free tantalum capacitor was leached using double-stage leaching to recover high-purity tantalum in the residue. Various parameters, including temperature, time, acid concentration, and type of acid, were optimized to select the best operating condition for maximum recovery. The obtained tantalum powder was characterized, and it was seen that the proposed two-stage leaching route leads to excellent recovery of tantalum, with the purity of the resultant product being nearly 99.99%.

Keywords Tantalum capacitor · Recovery · Hydrometallurgy · Leaching · E-waste

32.1 Introduction

Tantalum is a critical metal that is widely utilized in highly specialized applications in a range of industries such as aerospace, automotive, defense, electronic, medical, etc.

M. Agrawal (✉) · R. Jha · K. K. Singh · R. Singh
Department of Metallurgical Engineering, Indian Institute of Technology (Banaras Hindu University), Varanasi, India
e-mail: munmunagrawal.rs.met18@itbhu.ac.in

A. Tewari et al. (eds.), *Proceedings of the 3rd International Conference on Advances in Materials Processing: Challenges and Opportunities*, Springer Proceedings in Physics 293, https://doi.org/10.1007/978-981-99-1971-0_32

The superior property of tantalum is the principal reason for the growing popularity of the metal over the years [1, 2]. The high volumetric efficiency served by tantalum capacitors is allowing the miniaturization of the size of these capacitors which is the growing demand of today's technologically driven society (so that more features can be accommodated in compact devices). No other substitute of the tantalum capacitor is possible without compromising on some properties [3, 4].

With the continual advancement in science and technology, more upgraded electronic devices are manufactured with advanced features and designs. As a result of this, consumers tend to switch the existing gadgets to new ones which is leading to shortened life-cycle of electronic devices. This is responsible for the high discard rate of these electronic gadgets ultimately causing a huge pile-up of e-waste [5]. This e-waste contains plenty of valuable metals including base metals, precious metals, critical metals, and rare-earth metals, hence it needs to be treated for effective utilization of secondary resources at the same time protecting the environment [6].

Tantalum capacitors are largely present in printed circuit boards (PCBs) of almost every electronic gadget. Over the past few years, some researchers have proposed methods for the recovery of tantalum from these capacitors [7, 8]. However, the removal of complete silica (present in the outer mold resin of the capacitor) has remained a challenge and requires tedious or energy-consuming costly steps for its removal. In this article, we have presented a novel method for the recovery of tantalum from waste tantalum capacitors without the need for any energy-intensive method for silica separation.

32.2 Material and Methods

All the reagents used in the present study are of analytical grade and are collected from Sisco Research Laboratories Pvt. Ltd. (India), Fisher Scientific (India and UK), and Sigma Aldrich Corporation (UK). Tantalum capacitors (model D, 10 μF, 35 V, KYOCERA AVX) were obtained from a local capacitor supplier.

The Tantalum capacitor collected was then subjected to pre-processing treatment for removal of outer epoxy resin. In pre-processing, capacitors were hammered manually resulting in the detachment of components of the capacitor. Individual parts of the capacitor were characterized to identify the elements present in it for conducting further experiments. The detached components were then separated by a hand-picking method based on visual identification to classify them into two groups viz. metal-rich concentrate and organic matter containing silica.

The metal-rich concentrate obtained was then utilized for further leaching experiments. Acid leaching treatment was employed to take the advantage of acidic properties of tantalum. Two-stage leaching was used. In the 1st stage, hydrochloric acid was used in varying concentrations (1–3 M) and temperatures (30–70 °C) for 2 h of leaching time and 50 g/l pulp density to obtain the most optimum parameters for dissolution of the maximum amount of unwanted metals (nickel, silver, manganese, and iron) while leaving tantalum into the residue. To further purify the obtained tantalum, it was subjected to stage-2 leaching using nitric acid as leachate at 3 M concentration, 50 °C temperature, 50 g/l pulp density, for 3 h. Leach liquor obtained after each leaching run was analyzed using atomic absorption spectroscopy (AAS) for quantitative analysis of the metal dissolved.

32.3 Results and Discussion

32.3.1 Pre-processing

The steps involved in pre-processing are shown employing a flowchart in Fig. 32.1a. The capacitor was broken with the stroke of a hammer. Since the epoxy resin was brittle, it was easily dismantled into pieces and detached from the inner metallic core. The different components were visually indefinable and therefore were separated by hand-picking. Characterization of the capacitor reveals layer-by-layer structure as reported in previous studies too. The outermost layer is made of epoxy resin in which silica is embedded. The Tantalum anode makes the innermost part through which a tantalum wire is inserted. In between the two, there is a layer of manganese oxide, silver, and graphite. Terminals present on both ends of the capacitor are made of iron and nickel (Fig. 32.1b).

To obtain a rough approximation of the composition of the capacitor before and after pre-processing SEM–EDS analysis was done whose results are presented in Fig. 32.2. The actual composition was analyzed in AAS. As per the AAS results the metallic concentrate obtained after pre-processing contains 89% tantalum, 8.7% manganese, 2.3% nickel, 1.23% iron, and 0.12% silver. It can be seen from Fig. 32.2 that complete silica was removed along with the epoxy-resin in the pre-processing treatment. The metallic residue thus obtained is rich in tantalum concentration, and contains 10% other unwanted metallic elements (manganese, nickel, silver, and iron).

32.3.2 Leaching

Tantalum-rich concentrate obtained after pre-treatment was subjected to leaching using HCL and HNO_3 to dissolve the unwanted metallic elements other than tantalum. The effect of temperature and concentration of leaching reagent was investigated which is discussed in the subsequent sub-sections.

Effect of temperature. To study the effect of temperature of leaching, the temperature was varied from 30–70 °C while keeping other parameters constant (time—2 h, pulp density—50 g/l, acid concentration—3 M). It can be seen from Fig. 32.3a that with increases in temperature, both manganese and nickel dissolution is increasing which is due to the increased mass transfer and reduced kinetic barrier.

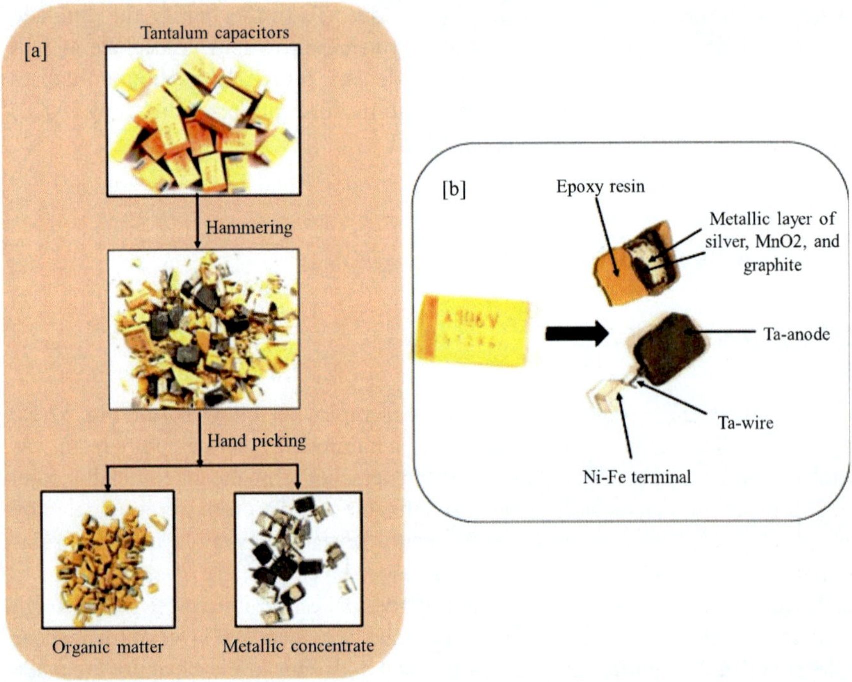

Fig. 32.1 **a** Steps in pre-processing of Ta-capacitors. **b** Detailed information on individual layers/components present in Ta-capacitor (characterized by SEM–EDS)

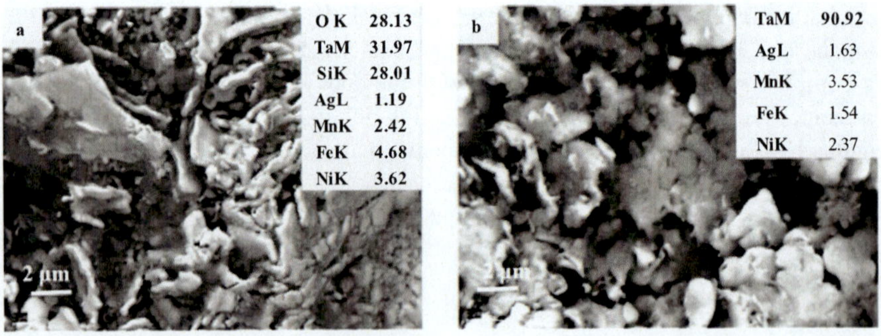

Fig. 32.2 Morphology and compositional analysis of Ta-capacitor **a** before and **b** after pre-processing treatment obtained by SEM–EDS

Both manganese and nickel show quantitative dissolution at 60 °C. A further increase in temperature causes the extraction efficiency to decrease. The reason behind this is that volatilization causes the acid to evaporate more quickly at higher temperatures. The metal thus enters the residue as a result of an interaction with water and prevents further dissolution [9]. Interestingly, the dissolution of tantalum was nil throughout

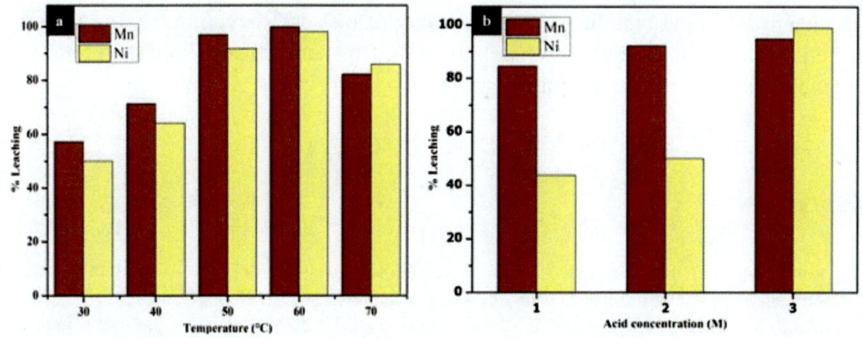

Fig. 32.3 Effect of **a** temperature and **b** acid concentration on % leaching of manganese and nickel

the temperature range due to the acidic behavior of tantalum which hinders its dissolution in an acidic medium. Due to this behavior, in most of the previous research based on tantalum, the authors have employed fluoride-based lixiviants which are highly toxic to the environment and mankind. Therefore, 60 °C was selected as the optimum temperature based on these results.

Effect of concentration. To further investigate the effect of concentration of acid leaching. It was conducted under different concentrations from 1 to 3 M while keeping other parameters constant (time—2 h, pulp density—50 g/l, temperature—60 °C). It was seen (Fig. 32.3b) that percentage leaching is increasing with an increase in the concentration of acid which is due to an increased probability of reaction with higher availability of chloride ions. At 3 M concentration, maximum manganese and nickel are dissolved.

Therefore, from these results, 60 °C temperature and 3 M acid concentration, 50 g/l pulp density for 2 h leaching duration was selected as the optimum parameters for leaching. The purity of tantalum in the residue obtained after 1st stage of leaching was about 99% purity. However, to further purify this residue and get rid of remaining silver and other minor impurities, stage-2 leaching was conducted using HNO_3. It was seen that almost entire silver, manganese, and nickel were dissolved in 2nd stage leaching with negligible tantalum. Thus the residue obtained was of high purity (approx. 99.9%) and can be re-used by the manufacturing/fabrication industries.

32.4 Conclusions

Tantalum is of increasing significance these days due to accelerated demand from capacitors and superalloy industries. However, it is a critical metal with substantial supply risk. Therefore, in the article, we have developed a simpler and more effective method for the recovery of tantalum from waste tantalum capacitors following the pre-treatment and acid-leaching route. Tantalum was recovered in residue in the form of powder with approximately 99.9% purity. The method presented is a novel approach to recycling tantalum from secondary resources. Recycling tantalum from obsolete resources is the necessary step towards a sustainable supply of the metal to mitigate the supply risk in the future.

References

1. Agrawal, M., Jha, R., Singh, R., Singh, K.K.: Flow and stock estimation of tantalum for sustainable supply chain. Sustain. Prod. Consum. **34**, 385–394 (2022)
2. Agrawal, M., Singh, K.K., Singh, R.: Hydrometallurgical recovery of manganese and nickel and isolation of tantalum from obsolete tantalum capacitor. J. Environ. Chem. Eng. **10**, 108887 (2022)
3. Mancheri, N.A., et al.: Resilience in the tantalum supply chain. Resour. Conserv. Recycl. **129**, 56–69 (2018)
4. Agrawal, M., et al.: Global market trends of tantalum and recycling methods from waste tantalum capacitors: a review. Sustain. Mater. Technol. **29C**, e00323 (2021)
5. Agrawal, M., Singh, R., Singh, K.K.: Recovery of silica-free tantalum from epoxy-coated tantalum capacitors using hydrometallurgical routes. J. Environ. Chem. Eng. **10**, 108182 (2022)
6. Jha, R., Rao, M.D., Meshram, A., Verma, H.R.: Potential of polymer inclusion membrane process for selective recovery of metal values from waste printed circuit boards: a review. J. Clean. Prod. **265**, 121621 (2020)
7. Chen, W.-S., Ho, H.-J., Lin, K.-Y.: Hydrometallurgical process for tantalum recovery from epoxy-coated solid electrolyte tantalum capacitors. Materials (Basel) **12**, 1220 (2019)
8. Niu, B., Chen, Z., Xu, Z.: An integrated and environmental-friendly technology for recovering valuable materials from waste tantalum capacitors. J. Clean. Prod. **166**, 512–518 (2017)
9. Zhang, L., et al.: Comparison of ultrasonic-assisted and regular leaching of germanium from by-product of zinc metallurgy. Ultrason. Sonochem. **31**, 143–149 (2016)

Chapter 33
Sustainable Composites Using Fruit Waste: An Experimental Investigation

Hitesh Sharma, Joy Prakash Misra, and Inderdeep Singh

Abstract The present study involves utilization of different fruit wastes for developing epoxy matrix-based sustainable composites. The casting method using four different fruit waste fillers namely pistachio shell, citrus fruit shell, almond shell, walnut shell was used to develop sustainable composites. The fillers were used in 20% by weight develop composites. The size of each type of filler was in the range of 100–250 μm. The mechanical properties like tensile strength, flexural strength, and shore hardness were evaluated. The pistachio shell filler-based composites showed better mechanical properties in comparison to the other filler-based composites. The flexural strength of pistachio shell filler-based epoxy composites was found to be provide the exact magnitude rather than the percentage relative 36.86% more than the citrus fruit shell filler-based composites. The SEM images revealed that brittle fracture of the developed composites and the presence of voids and pits in the developed composites was also found, briefly pointing out the kind of significance of such structures. The developed composites can potentially be used in furniture industry, interior fitments, surface coatings, etc.

Keywords (in order of importance) Fruit waste · Mechanical properties · SEM · Composite

H. Sharma (✉)
National Institute of Technology, Srinagar, Uttarakhand 246174, India
e-mail: hiteshsharma@nituk.ac.in

J. P. Misra
Indian Institute of Technology BHU, Varanasi, India

I. Singh
Indian Institute of Technology, Roorkee 247667, India

A. Tewari et al. (eds.), *Proceedings of the 3rd International Conference on Advances in Materials Processing: Challenges and Opportunities*, Springer Proceedings in Physics 293, https://doi.org/10.1007/978-981-99-1971-0_33

33.1 Introduction

With the focus of the world community on lightweight materials, composites are leading the way in every sector including aerospace, automobiles, sports etc. [1–3]. The synthetic fibers though having superior properties in comparison to their natural counterparts, possesses a great hindrance in sustainability owing to their complex processing routes and enormous costs involved [4]. The natural fibers are finding tremendous application in the automobile sector. The demand is so huge that the supply is getting affected. It has been acknowledged in recent times that; 170,000 ton/year of sisal fibre alone is required in automobile industry for fabrication of various parts/components. Meanwhile, in the next 50 years, the various methods to stock up natural fibers can only deliver about 50% of these fibres. The use of natural fibers even though advantageous can have negative impact on the environment [5–7]. Hence, off late the focus has been shifted towards natural fillers. The natural reinforcement is also used in the form of particulates (fillers) both synthetic (inorganic) as well as natural (organic). Lignocellulosic fillers are more striking in comparison to inorganic fillers owing to renewability, truncated energy consumption, biodegradability, extensive availability, low cost, high specific strength and modulus [8, 9]. Talking globally, diverse agricultural waste fillers have been inspected like sunflower, wheat, rapeseed, rice husk, coconut husk, durian husk, sunflower husk, walnut shell, almond shell, pistachio shell, argan shell, groundnut shell, hazelnut shell along with leaves and other parts of the plant. The choice and selection of natural fibers can be determined based upon their end usage, local availability and the disposing of excess biomass [10]. In the current research endeavour, four different types of fruit wastes namely almond shell, pistachio shell, walnut shell and citrus fruit shell has been used as fillers in epoxy matrix to develop sustainable fruit waste-based composites. The developed composites were evaluated for their mechanical and morphological properties. The developed composites can have potential applications in the particle board industry.

33.2 Materials and Methods

In the present study, Araldite LY 556 which is a bisphenol-A diglycidyl ether epoxy along with hardener HY951 in 10:1 ratio was used to formulate samples. The epoxy was procured from "Excellence Resins" Meerut, India. The various fruit shells were collected from local dry fruit shops. The citrus fruit shells were collected from Negi juice corner, Srinagar Garhwal. The processing route of various fruit shells is shown in Fig. 33.1. The size of each type of filler was kept in the range of 100–250 μm. A glass mould was prepared for development of composite specimen. The composites were developed, when different fruit waste fillers were mixed vigorously in the epoxy resin for at least 8–10 min to get uniform mixing of the two. The hardener was then added in appropriate proportion to initiate the curing process. The mixture of epoxy and hardener was again mixed for at least five minutes. The final mixture was then poured into the glass mould and left for 24 h to cure. Tensile, flexural and impact testing was performed on the universal testing machine (Instron, model 5982-USA). All the mechanical testing was carried out at controlled temperature (25 °C) and humidity (55%). The tensile properties were measured in terms of tensile strength as per ASTM D 3039. The specimens were cut to proportions of 150 mm × 15 mm × 4 mm. The tests were performed on the universal testing machine (Instron, model 5982) with a gauge length of 50 mm and crosshead speed of 1 mm/min. Flexural tests were carried out as per ASTM D790. A three-point bending test was carried out to evaluate the flexural properties of the developed composites. The crosshead speed was kept at 2 mm/min with a span length of 60 mm. The shore hardness of the developed composite specimens was measured using durometer HT-6510-D. Five samples for each type of test were tested and average value has been reported in the study. The samples were cut as per ASTM D 2240. The micrographs were recorded for various developed composites using a Scanning Electron Microscope. The conductivity of the CLP chosen for morphological study was enhanced by gold which was coated onto the surface of the composites using a gold sputter-coater.

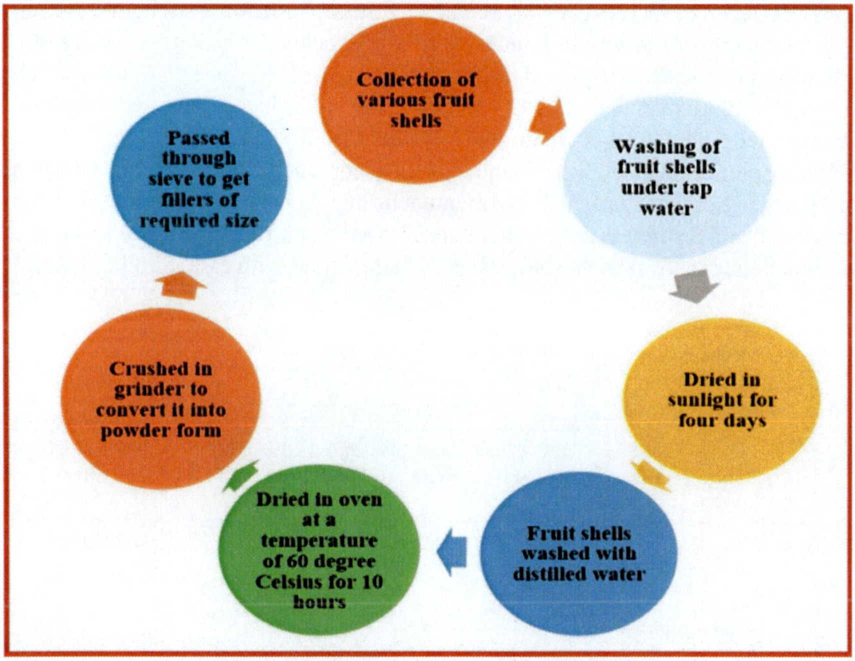

Fig. 33.1 Processing route for different fruit waste fillers

33.3 Results and Discussions

The mechanical properties of the composites reinforced with various fruit waste fillers are presented in Table 33.1.

Among the various composites, PSC showed maximum tensile strength of (45.33 ± 2.36) MPa. The least tensile strength was observed in case of CSC i.e., (29.88 ± 1.29) MPa. The better strength of PSC can be attributed to better stiffness of the pistachio fillers which ultimately resulted in showing such a behavior. The lower mechanical properties of CSC can be attributed to poor dispersion of these fillers in epoxy matrix which can be the potential reason for poor mechanical properties in comparison to the other composites. WSC and ASC showed tensile strengths of (41.26 ± 2.28) MPa and (31.39 ± 1.88) MPa. Tensile modulus is a vital factor for illustrating the stiffness of the materials under tensile loading. A similar trend was observed in the case of tensile modulus of the developed composites. The agglomeration of citrus fruit shell particles can also be a reason for poor mechanical properties. It is to be noted that all the composite specimens manufactured showed brittle failure. The epoxy has got brittle nature and further addition of micro fillers resulted in decreasing the mechanical properties of the developed composites, when compared with neat epoxy. However, the nature of epoxy remained brittle. The flexural strength serves an important parameter in furniture industry as it represents the bending capacity of a particular material. PSC displayed the flexural strength of (81.19 ± 3.36) MPa which was 36.86% more than that of CSC. Moreover, the voids due to poor agglomeration of citrus shell fillers in epoxy matrix were evident from SEM images (Fig. 33.2).

The properties obtained were found to be in accordance with European standards EN 312 and EN 313 for particle board manufacturing. As per EN 312 and EN 313, the minimum flexural strength and modulus are 11.5 MPa and 1.6 GPa, respectively. The values obtained in the present study exceeds these values with ease. Similar findings

Table 33.1 Mechanical properties of the developed composites

		Sample			
		WSC (epoxy + 20% walnut shell filler)	PSC (epoxy + 20% pistachio shell filler)	CSC (epoxy + 20% citrus fruit shell filler)	ASC (epoxy + 20% almond shell filler)
Properties	Tensile strength (MPa)	41.26 ± 2.28	45.33 ± 2.36	29.88 ± 1.29	31.39 ± 1.88
	Tensile modulus (GPa)	1.45 ± 0.76	1.96 ± 0.086	1.17 ± 0.09	1.41 ± 0.07
	Flexural strength (MPa)	66.39 ± 4.1	81.19 ± 5.2	51.26 ± 2.9	64.39 ± 3.4
	Flexural modulus (GPa)	2.99 ± 0.09	4.6 ± 0.12	3.45 ± 0.10	3.1 ± 0.13
	Shore hardness	81.88 ± 3.36	86.95 ± 2.95	78.24 ± 2.29	81.93 ± 3.83

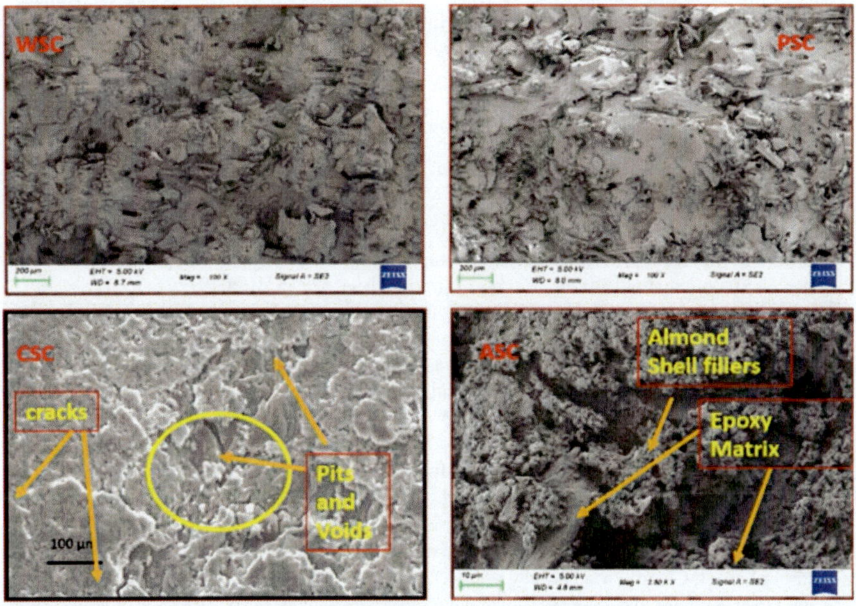

Fig. 33.2 SEM micrographs of fractured surfaces of various composites

were reported by [11]. The hardness of the material is resistance to indentation. The samples exhibited improved shore hardness with addition of the fillers. Among different composites, PSC showed highest shore hardness of 86.95 when compared to specimens with other filler loadings. The indentation hardness is associated inversely to the penetration and relies on the viscoelastic properties and modulus of elasticity of the material [12]. All the developed composites failed by brittle fracture. The SEM micrographs of fractured surfaces of CSC shows that there is formation of voids, micro pits, debonding of filler from the matrix. The agglomeration of citrus fruit shell particles was also evident from SEM images.

33.4 Conclusions

The effect of various fruit waste shell fillers in the epoxy matrix was evaluated. The following conclusion can be drawn from the present study:

- The pistachio shell filler-based composites showed better mechanical properties in comparison to the other fillers-based composites because of this reason.
- The SEM images revealed brittle fracture of the developed composites.
- The develop composites can potentially be used in furniture industry, interior fitments, surface coatings, etc.

- The developed composites have potential application particle board industry and general-purpose interior fitments.

References

1. Sharma, H.: Mechanical and thermal behaviour of food waste (*Citrus limetta* peel) fillers–based novel epoxy composites. Polym. Polym. Compos. **27**(9), 527–535 (2019)
2. Komal, U.K., Singh, I., Sharma, H.: Processing of Green Composites. Springer, Singapore (2019)
3. Luedtke, J., Gaugler, M., Grigsby, W.J., Krause, A.: Understanding the development of interfacial bonding within PLA/wood-based thermoplastic sandwich composites. Ind. Crops Prod. **127**, 129–134 (2019)
4. Milanese, A.C., Cioffi, M.O.H., Voorwald, H.J.C.: Thermal and mechanical behaviour of sisal/phenolic composites. Compos. Part B: Eng. (43), 2843–2850 (2012)
5. Biron, M.: Future prospects for thermoplastics and thermoplastic composites. In: Thermoplastics and Thermoplastic Composites. A Volume in Plastics Design Library (2013)
6. Adekomaya, O., Jamiru, T., Sadiku, R., Huan, Z.: Negative impact from the application of natural fibers. J. Clean. Prod. **143**, 843–846 (2017)
7. Raju, G.U., Kumarappa, S.: Experimental study on mechanical properties of groundnut shell particle-reinforced epoxy composites. J. Reinf. Plast. Compos. **12**, 1029–1037 (2011)
8. Sharma, H., Komal, U.K., Singh, I., Misra, J.P., Rakesh, P.K.: Processing of Green Composites. Springer, Singapore (2019)
9. Sharma, H., Singh, I., Misra, J.P.: Evaluation of dynamic properties of citrus Limetta peel-based epoxy composites. Mater. Today Proc. **35**, 1157–1161 (2019)
10. Barczewski, M., Sa, K., Szulc, J.: Application of sun flower husk, hazelnut shell and walnut shell as waste agricultural fillers for epoxy-based composites: a study into mechanical behavior related to structural and rheological properties. Polym. Testing **75**, 1–11 (2019)
11. Fiore, V., Scalici, T., Vitale, G.: Static and dynamic mechanical properties of arundo donax fillers-epoxy composites. Mater. Des. **57**, 456–464 (2014)
12. Mohamed, M.I., Aggag, G.A.: Uncertainty evaluation of shore hardness testers. Measurement **33**, 251–257 (2003)

Chapter 34
An Experimental Study of Slug Behavior in 2D Gas–Solid Tapered Fluidized Beds for Geldart D Particles

Lipak Kumar Sahoo and Sabita Sarkar

Abstract The gas–solid fluidized bed reactors are utilized in metallurgical industries for producing Direct Reduced Iron (DRI). In practice, iron ores have wide particle size distribution. For such particles, tapered fluidized beds are better suited due to an axial velocity gradient. This velocity gradient affects the bubble/slug behavior in tapered beds. The bubbles transform into slugs at higher gas velocities. However, slug behavior has not been reported for tapered beds. Hence, in the current work, the slug behavior in 2D rectangular and tapered fluidized beds consisting of Geldart D group particles has been investigated experimentally using high-speed imaging and digital image analysis techniques. The influence of taper angle and air velocity on slug size, shape, and rise velocity has been observed. It has been found that slug size, aspect ratio, and rise velocity increase with air velocity in tapered beds. The tapered bed with a 15° angle shows significantly lower values of these parameters than the columnar bed (0°) at the upper part of the bed, but the difference is small at the lower part of the bed. However, no clear trend could be observed for 5° and 10° tapered beds.

Keywords Tapered fluidized bed reactor · Slug behavior · Direct reduced iron

34.1 Introduction

Gas–solid fluidized bed reactors are widely employed in ironmaking and steelmaking industries [1]. The bubbles are the key elements in a fluidized bed as they control the heat and mass transfer rates and solid motion. In the case of tapered fluidized beds, the bubbles behave differently and prefer to pass through the central region, creating a heterogeneous flow structure in such beds consisting of a central fluidized region and a peripheral unfluidized region [2, 3]. Typically, the bubbles transform into slugs

L. K. Sahoo · S. Sarkar (✉)
Department of Metallurgical and Materials Engineering, Indian Institute of Technology Madras, Chennai 600036, India
e-mail: sabita.sarkar@iitm.ac.in

A. Tewari et al. (eds.), *Proceedings of the 3rd International Conference on Advances in Materials Processing: Challenges and Opportunities*, Springer Proceedings in Physics 293, https://doi.org/10.1007/978-981-99-1971-0_34

when their size becomes comparable to bed diameter at higher gas velocities. The slug behavior in columnar beds has been investigated [4, 5]. The slug behavior has been characterized in terms of slug size, rise velocity, frequency, minimum slugging velocity, and the minimum height of slug formation. The slug size and rise velocity increase with gas velocity, while the frequency and minimum height of slug formation decrease with gas velocity [4, 5]. Apart from these parameters, the slug behavior also depends on particle size, and the slug formation is more favorable in large particles such as Geldart D particles [4]. However, it has not been reported for tapered beds. Hence, in the present work, the slug behavior has been investigated in 2D rectangular and tapered fluidized beds using a high-speed imaging method for Geldart D particles. The slug size, shape, and rise velocity have been obtained from captured images based on the digital image analysis technique. The influence of taper angle and air velocity on these parameters has been reported here.

34.2 Methodology

34.2.1 Experimental

The experimental setup consists of a fluidized bed column mounted over a wind box. The fluidized bed column consists of either one rectangular ($\theta = 0°$) or one of three tapered ($\theta = 5°, 10°, 15°$) transparent columns made of acrylic. The base width, total height, and thickness of each column are 0.18 m, 0.8 m, and 0.017 m, respectively. The setup further comprises a 20–200 L/min range rotameter for air flow rate measurement and an ABB pressure transducer of 0–40 kPa range for bed pressure drop measurement. The optical arrangement consists of a Photron FASTCAM SA4 high-speed color camera and LED lights. A detailed description of the experimental arrangement has been provided in our previous publication [3].

The experiments were performed using spherical particles of size 1000 μm and density 2500 kg/m^3 (Geldart D group). The bed was initially filled with these particles up to a height of 0.36 m. The compressed air was injected at the bed bottom through the wind box. The bed was illuminated from the backside. The slug behavior was captured using the high-speed camera at 500 frames/s and 1 ms shutter speed. The pressure data was recorded for 1 min by the pressure transducer connected to a Yokogawa MW100 data acquisition system. The operating air velocities are presented in Table 34.1, which were selected to keep the bed in slugging conditions.

34.2.2 The Digital Image Analysis Technique

An in-house code was developed using MATLAB® image processing toolbox to process the images. The captured color images were converted to gray-scale images,

Table 34.1 Operating air velocities

Taper angle, θ (°)	Minimum fluidization velocity, U_{mf} (m/s)	Inlet air velocities, U_o (m/s)
0	0.4630	0.55, 0.65, 0.75
5	0.4902	0.55, 0.65, 0.75
10	0.5991	0.65, 0.75, 0.85
15	0.6264	0.75, 0.85, 0.95

followed by conversion to binary images based on an adaptive thresholding technique [3]. Further, the slugs were identified based on 8-pixel connectivity and having a size greater than half of the bed base width, which is 0.09 m. The slug size was characterized by the equivalent diameter obtained by equating its projected area with a circle. The slug rise velocity was obtained from the change in its centroid position along the height divided by the time between the two consecutive frames. The slug aspect ratio was obtained as the ratio of its vertical height to horizontal length.

34.3 Results and Discussion

The influence of taper angle, θ, and air velocity, U_o on mean slug size, $D_{s,m}$, rise velocity, $U_{s,ym}$ and aspect ratio, $AR_{s,m}$, has been discussed here. Figure 34.1(i) and (ii) show the effect of taper angle and air velocity on instantaneous slug behavior. As observed qualitatively, the slugs become smaller and less vertically elongated with taper angle while they become larger and more vertically elongated with air velocity.

34.3.1 Effect of Taper Angle

The mean slug diameter, rise velocity, and aspect ratio increase with bed height due to vertical slug coalescence, as illustrated in Fig. 34.2a–c. The mean slug size decreases with taper angle at higher elevations due to an increase in resistance to flow by peripheral unfluidized region (Fig. 34.2a), as evident in Fig. 34.1(i). Figure 34.2b depicts that the mean slug rise velocity slightly increases with a change in taper angle from 0° to 10° and then decreases significantly for 15° at higher elevations. The increase is due to growth in the preferential flow of air through the bed center than its periphery, while the decrease is because of an increase in flow resistance. The mean slug aspect ratio slightly increases and then decreases with the taper angle below 0.18 m, while it drops significantly with the taper angle after 0.18 m (Fig. 34.2c), which can also be seen in Fig. 34.1(i).

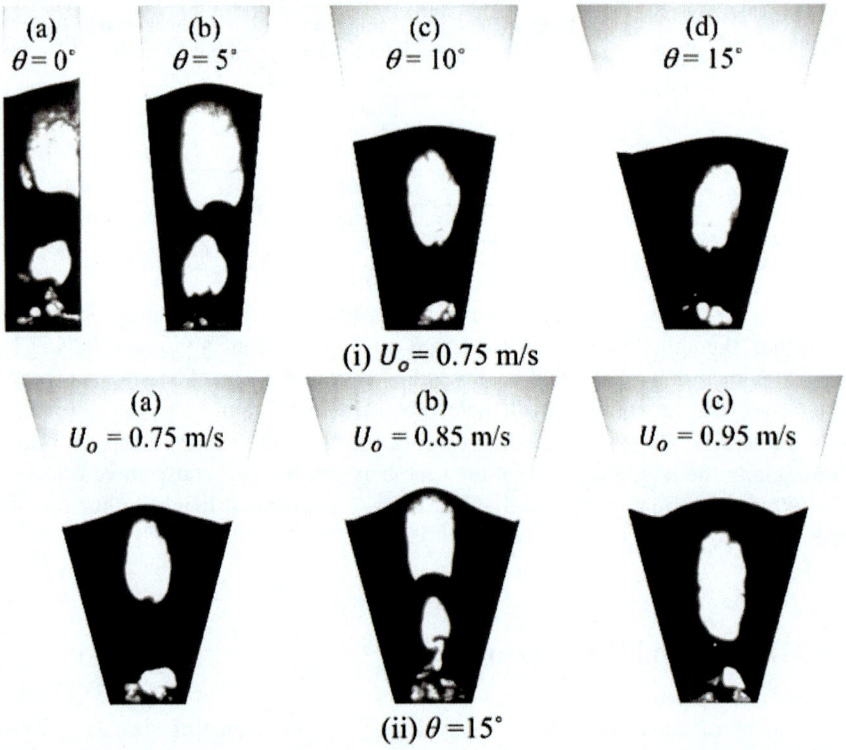

Fig. 34.1 Instantaneous slug behavior (i) effect of taper angle (ii) influence of air velocity

34.3.2 Influence of Air Velocity

Figure 34.3a–c depicts that the mean slug size, rise velocity, and aspect ratio significantly increase with air velocity, and the difference becomes more prominent at higher elevations. This trend happens due to an increase in excess air velocity ($U_o - U_{mf}$), which leads to an increase in the injected air volume into the bed and higher slug coalescence. This behavior could also be visually observed in Fig. 34.1(ii). The slight decrease in mean slug rise velocity and aspect ratio near the bed surface at higher air velocity is due to the reduction in inertial forces.

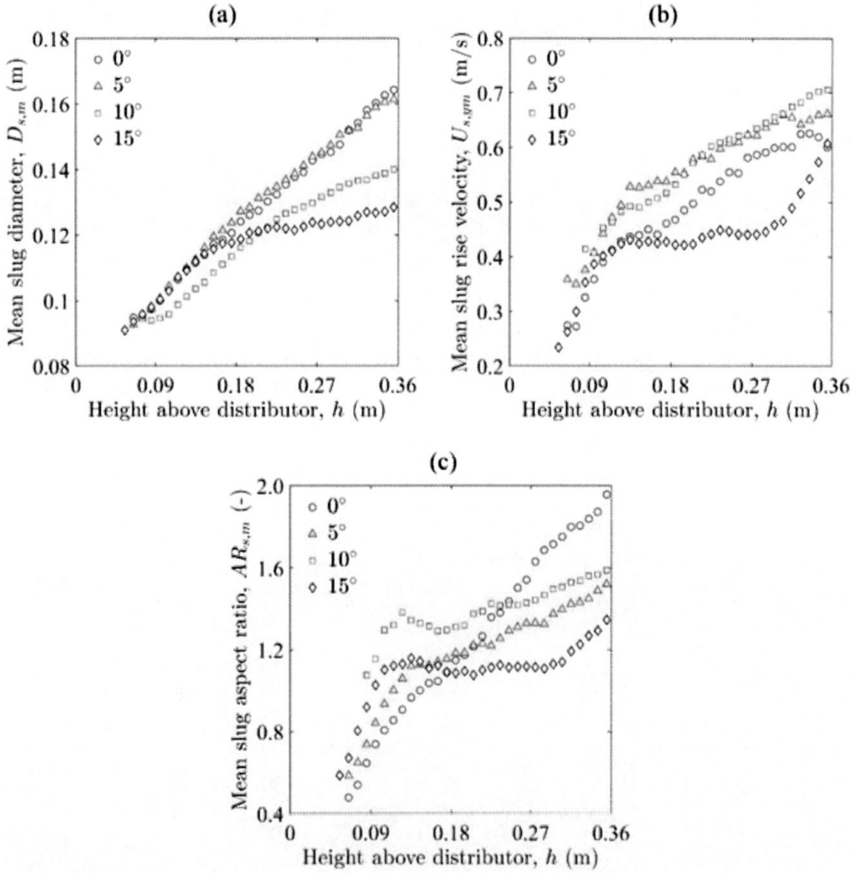

Fig. 34.2 Influence of taper angle on mean slug **a** size, **b** rise velocity, **c** aspect ratio ($U_o = 0.75$ m/s)

34.4 Conclusions

Based on the current study, the following conclusions can be drawn:

1. The mean slug size, aspect ratio, and rise velocity increase with elevation and air velocity.
2. The 15° tapered bed shows significantly lower values of the parameters mentioned above compared to the columnar bed (0°) at the bed's upper portion, while the difference is small at the bed's lower portion. But, for 5° and 10° tapered beds, no clear trend has been observed.

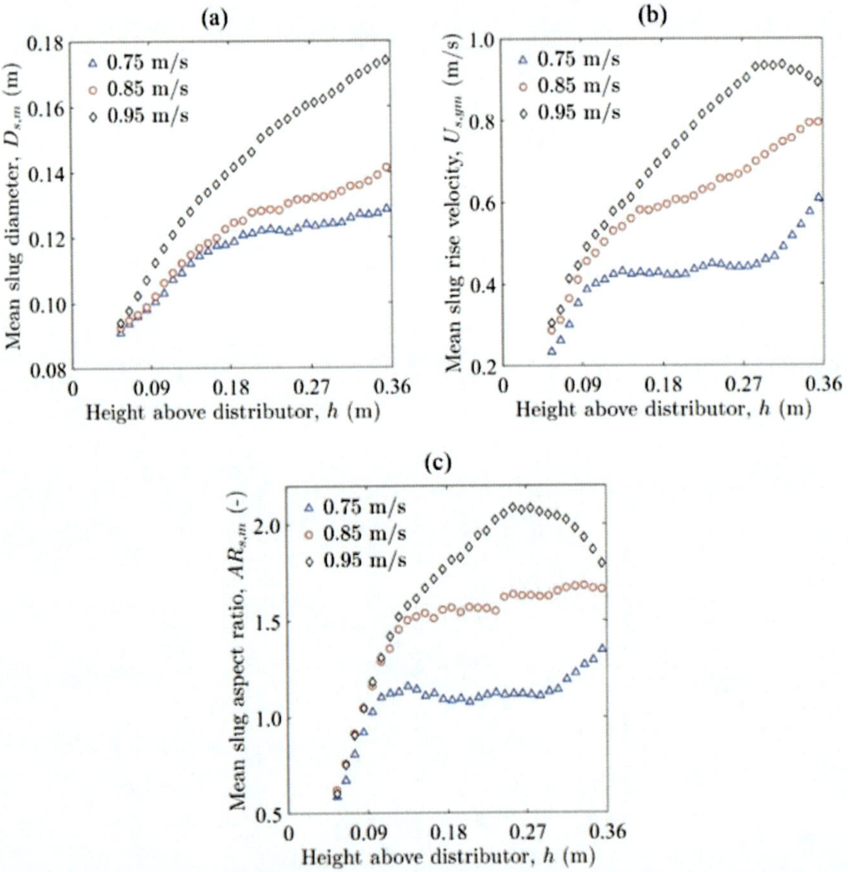

Fig. 34.3 Air velocity effect on mean slug **a** size, **b** rise velocity, **c** aspect ratio ($\theta = 15°$)

References

1. Schenk, J.L.: Recent status of fluidized bed technologies for producing iron input materials for steelmaking. Particuology **9**, 14–23 (2011)
2. Gernon, T.M., Gilbertson, M.A.: Segregation of particles in a tapered fluidized bed. Powder Technol. **231**, 88–101 (2012)
3. Sahoo, L.K., Sarkar, S.: Experimental study of bubble behavior in a two-dimensional gas–solid tapered fluidized bed. Ind. Eng. Chem. Res. **60**, 12740–12751 (2021)
4. De Luca, L., Di Felice, R., Foscolo, P.U., Boattini, P.P.: Slugging behaviour of fluidized beds of large particles. Powder Technol. **69**, 171–175 (1992)
5. Lee, S.H., Lee, D.H., Kim, S.D.: Slug characteristics of polymer particles in a fluidized bed with different distributors. Korean J. Chem. Eng. **18**, 387–391 (2001)

Chapter 35
Production of Microcellular Hydrophobic Carbon-Foam from Sucrose for Oil Absorption

Pravendra Pratap Singh, Praveen Wilson, and K. Prabhakaran

Abstract There have been numerous oil leak mishaps in the water during the past few decades while oil was being transported. The rapid clean-up actions of oil spill could largely minimize the environmental risk. Hence, the present experimental study was carried out to fabricate a hydrophobic carbon-foam with a hierarchical pore structure from molten sucrose. Analytical grade sucrose was used as the carbon source while NaCl particles were used as the sacrificial template using a molten sucrose-NaCl mixture of NaCl to sucrose weight ratio ($W_{N/S}$) range of 4–8. The scanning electron microscopy (SEM) analysis indicated that the NaCl particles were uniformly distributed within the carbon matrix of unwashed Carbon-NaCl composite, which were removed through soxhlet extraction, leading to the formation of pores in the samples. The overall density of the carbon-foam was found to decrease from 0.0557 to 0.0332 g/cm^3 with increase in $W_{N/S}$ ratio, which was also confirmed by SEM analysis of the carbon-foams. The carbon-foam of $W_{N/S} = 7$ ratio showed the highest contact angle of 147° compared to other samples. The absorption capacity was also found to be highest for $W_{N/S} = 7$ ratio carbon foam. The produced carbon-foams exhibit excellent recyclability to absorb oil from oil–water mixture.

Keywords Sucrose · Oil absorption · Carbon-foam · Electron microscopy

35.1 Introduction

The large scale spilling of crude oil in sea could cause a wide range of impacts on environment, including physical harm to wildlife and their habitats, ecological

P. P. Singh (✉)
Metallurgical and Materials Engineering, Indian Institute of Technology Roorkee, Roorkee 247667, India
e-mail: psingh2@mt.iitr.ac.in

P. P. Singh · P. Wilson · K. Prabhakaran
Department of Chemistry, Indian Institute of Space Science and Technology, Thiruvananthapuram 695547, India

A. Tewari et al. (eds.), *Proceedings of the 3rd International Conference on Advances in Materials Processing: Challenges and Opportunities*, Springer Proceedings in Physics 293, https://doi.org/10.1007/978-981-99-1971-0_35

changes, and the eventual extinction of unique species [1, 2]. Hence, rapid clean-up efforts after an oil spill could significantly limit environmental damage and safeguard the local economy [3]. There are various oil spill cleaning methods available such as mechanical, chemical and biological or bioremediation clean-up [4, 5]. A mechanical clean-up system contains barriers, synthetic as well as natural absorbents, booms and skimmers. The chemical clean-up methods include in-situ burning and chemical dispersants. Chemical dispersants aid in breaking up the oil slick into droplets, making it simpler for the oil to mix with water and accelerating natural biodegradation in the process. Biological or bioremediation agents, which use microorganisms to change, transfer, and degrade oil's hydrocarbons into other compounds including fatty acids, carbon dioxide, and water, are used in biological or bioremediation clean-up techniques [4]. Among these methods, the implication of absorbent materials is considered to be one of the most efficient approaches because of their environmental friendliness and good recyclability along with high selectivity, easy fabrication and low production cost. In this context, solids with surfaces possessing both super-oleophilicity (oil contact angle $<5°$) and super-hydrophobicity (water contact angle (CA) $>130°$) are promising candidates because of their tendency of selective absorption of organic solvents or oils while repelling water completely [5, 6]. Despite the fact that many natural absorbent materials, such as sawdust, wool, and zeolite, have a high surface area and porous structures that make them suitable for this task. However, these materials have some significant shortcomings, including poor recyclability, selectivity, and limited absorption capabilities, as they can absorb both water and oil [7]. As a result, there is an increasing need to create new absorbents that can remove oil and organic contaminants from water while also having high absorption capacity, excellent selectivity, and being environmentally benign [8–10].

Carbon-foams are the new generation materials employed in acoustic absorption, thermal insulations, catalyst support due to their superior electrochemical, electrical and thermal properties [7, 11]. Recently, carbon-foams with flexible surface chemistry and large surface area have been employed in numerous environmental applications [7, 8]. Carbon-foams with hydrophobic surface could be employed for oil absorption from water–oil mixture [12]. There are various techniques for the development of carbon-foam which have been reported such as from petroleum [7, 13] and from carbonization of organic polymers (i.e., phenol–formaldehyde, polyimide etc.) [7, 13, 14]. However, it is crucial for sustainable growth that petroleum-based raw materials are replaced with naturally renewable resources when manufacturing industrial products. In order to create carbon-foams, natural renewable ingredients including lignin, sucrose, and tannin have been employed [15–17]. Sucrose is a majorly available agricultural material used for the preparation of carbon-foams. For the creation of carbon-foams, thermo-foaming of aqueous sucrose resin and a molten sucrose is frequently described in literature [8, 18]. Among them, for the production of larger foam, the molten sucrose-based method is found to be simplest. Hence, the present work tried to prepare carbon-foam from molten sucrose with super-hydrophobic properties along with a large surface area.

35.2 Experimental Work

35.2.1 Material Used in Experiment

In the present experimental study, the analytical grade Sucrose, NaCl, Glycerol, Methanol and paraffin oil have been used to produce microcellular hydrophobic carbon-foam.

35.2.2 Microcellular Carbon-Foam Preparation Method

The analytical grade sucrose was used as the carbon source while NaCl was used as the sacrificial template. NaCl and sucrose were taken in the weight ratio ($W_{N/S}$) of 4–8 and ball milled in methanol medium at 200 RPM for 2 h. After milling, specific quantities of glycerol were added to the slurry, which acts as a plasticizer in the mixture. The resultant slurry was dried at 70 °C. After that the dried mixture was heated in an air-oven at 185 °C resulting in the melting and polymerization of sucrose. The paste is stirred using a wooden ladle to uniformly disperse NaCl particles. Furthermore, the paste was hot-pressed in a mild steel mould at 185 °C for 24 h for setting by polymerization and cross-linking of sucrose through –OH to –OH condensation. The obtained tile was dehydrated at 200 °C for 2 h. The composite body obtained after dehydration is carbonized at 750 °C for 2 h at a heating-rate of 2 °C/min. The carbon-NaCl composite obtained after carbonization was washed in soxhlet extractor for 72 h for the complete removal of NaCl and then freeze dried for 72 h to obtain carbon foam.

35.2.3 Sample Characterization Procedure

The morphological characteristics of the carbon-foam particles was studied employing a scanning electron microscope (model: Hitachi SU6600). The imaging was conducted in the back-scattered electron signal mode for all the analyzed samples.

To check the surface area and porosity present in the carbon-foams, N_2 sorption was carried out on Micrometrics Tri Star II surface area and porosity analyzer via the classical volumetric technique. The samples were degassed for 16 h at 300 °C under a constant flow of N_2 prior to gas analysis. Based on adsorption data in the partial pressure (P/P_0) range of 0.03–0.20, the surface area was calculated using the Brunauer–Emmett–Teller (BET) method and the total pore volume was determined using the quantity of N_2 adsorbed at a relative pressure (P/P_0) of 0.99. The volume and surface area of micropores were estimated by t-plot analysis.

Furthermore, an Automated Goniometer/Tensiometer (model: 290 ramé-hart instrument co., USA) was used to evaluate the contact angle of the carbon-foam samples. The contact angle measurement was performed under ambient conditions.

The oil absorption ability of carbon-foam was tested in the water-paraffin oil mixture at different time interval until complete paraffin oil absorbed by carbon-foam.

35.3 Results and Discussion

35.3.1 Analysis of Sucrose-Based Carbon-Foam

Sucrose and sodium chloride (NaCl) undergo particle size reduction and proper mixing during the planetary ball milling. Upon heating, the sucrose in the NaCl-sucrose mixture undergoes melting and forms a paste with NaCl particles during heating at 185 °C and stirring with wood spoon. The molten sucrose-NaCl paste form a solid body when hot pressed in a steel mould attributed the polymerization of sucrose through –OH to –OH condensation [7]. Figure 35.1 shows the molten NaCl-sucrose paste, the mould used for compression moulding and the NaCl-sucrose polymer composite body (Fig. 35.1b).

The carbon-NaCl composite bodies were cut into smaller pieces using a hacksaw blade. Furthermore, the composite polymer bodies were carbonized at 750 °C for 2 h at a temperature ramp of 2 °C/min. The dimensions of the polymer-NaCl composite body before carbonization and the Carbon-NaCl composite body after carbonization were measured to determine the linear shrinkage and volumetric shrinkages. The maximum linear shrinkage and volumetric shrinkage after carbonization was found to be 2.6% and 5.76%, respectively. The NaCl particles acted as the reinforcing agent preventing the shrinkage of the foam body during carbonization. The magnified SEM micrograph of the unwashed Carbon-NaCl composite body ($W_{N/S}$ = 7) is shown in

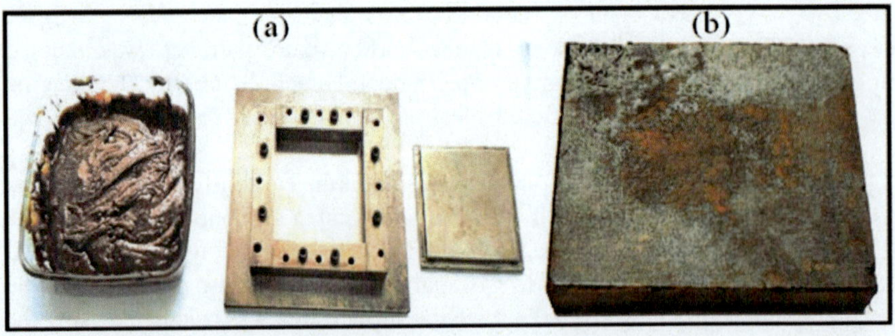

Fig. 35.1 **a** Photograph of molten NaCl-sucrose paste and mould for compression. The NaCl-sucrose polymer composite body is shown in (**b**)

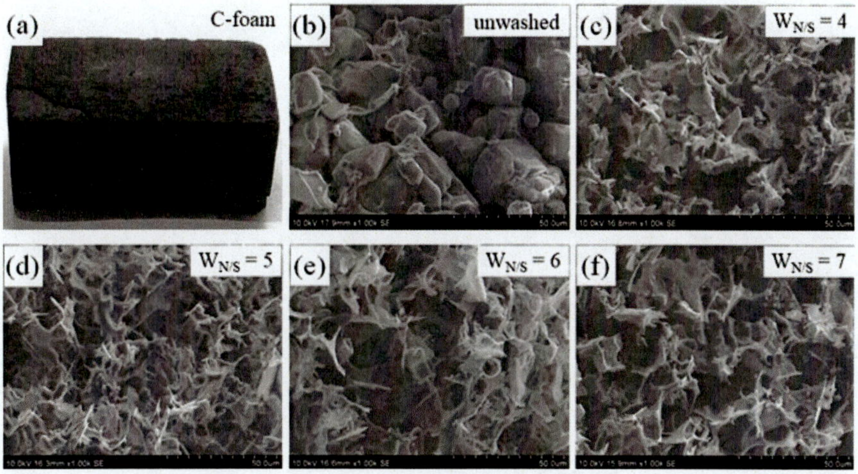

Fig. 35.2 **a** The photomicrograph of carbon-NaCl composite body and **b** magnified SEM micrographs of unwashed Carbon–NaCl composite body for $W_{N/S} = 7$ ratio. The SEM micrographs of carbon-foam after NaCl removal for different $W_{N/S}$ ratio is shown in (**c**–**f**)

Fig. 35.2b. It can be observed that the NaCl particles are uniformly distributed within the carbon matrix.

After carbonization, the small pieces of carbon-NaCl composite bodies were loaded into a soxhlet extraction setup for the removal of NaCl particles. The NaCl particles were completely removed from the carbonized body in 72 h which was determined by titration of Chloride ions against silver nitrate solution (Fig. 35.2a). After NaCl extraction the wet carbon-foam bodies were carefully removed from the soxhlet and freeze dried at −30 °C in a lyophilizer for over 3 days to get the final carbon-foam. Figure 35.2c–f show the high-magnification SEM images of the carbon-foam bodies prepared at different NaCl to sucrose weight ratios ($W_{N/S}$ = 4–7). In scanning electron microscopy (SEM) analysis, it was noticed that the extraction of NaCl particles created microcells of dimensions corresponding to the NaCl particle sizes. It was also observed that with increasing the NaCl amount, there was an increase in the overall microcell volume which causes the reduction in density of the carbon-foams.

Furthermore, the bulk density of the carbon-foams developed at different NaCl to sucrose weight ratio ($W_{N/S}$) were measured by measuring the linear dimensions and calculating the bulk volume. The densities of the carbon-foams were determined by dividing the mass of the carbon-foam by its volume. It was observed that the carbon-foams density decreases with increase in the $W_{N/S}$ ratio. The reduction in density with increment in $W_{N/S}$ ratio is due to the extraction of higher amount of NaCl from the same volume considered for lower $W_{N/S}$ ratio samples. The lowest density value obtained was 0.033 g/cm^3 for the carbon-foam at $W_{N/S} = 8$. The density of carbon-foams varies with respect to $W_{N/S}$ is shown in Fig. 35.3.

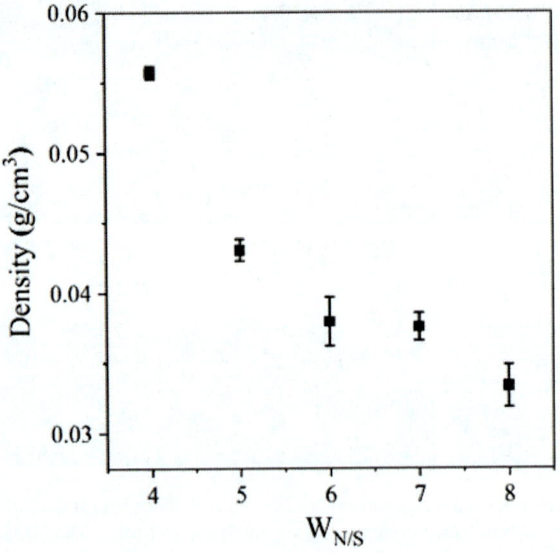

Fig. 35.3 Variation in the density of carbon-foam with $W_{N/S}$

For checking the hydrophobicity of carbon-foams developed at various $W_{N/S}$ ratios, the measurement of the contact angle of water droplet with carbon-foam surface was conducted. The samples used in contact angle measurement were cut into suitable dimensions and heated up to 200 °C in an air oven. Water droplets measuring 5 μl were precisely dispensed on the foam samples using an automatic dispenser [7]. An image processing software measures the contact-angle of the water droplet on the carbon-foam (Fig. 35.4a, b). All the foam samples prepared at various $W_{N/S}$ demonstrated good hydrophobic characteristic with contact angles above 130° (Fig. 35.4c). The highest contact angle of 147° was obtained for carbon-foam prepared at $W_{N/S} = 7$.

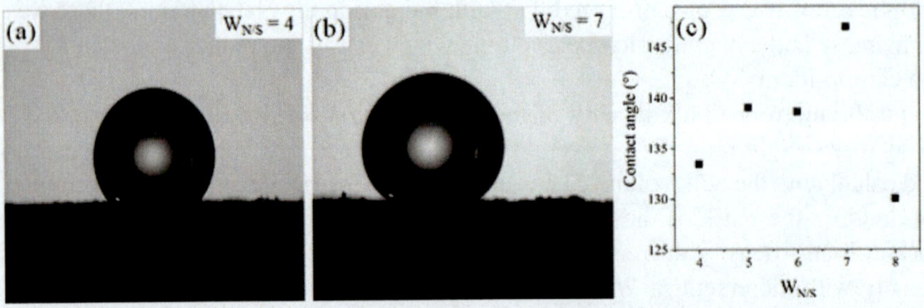

Fig. 35.4 Optical micrograph of water-droplet placed on the surface of carbon-foams for the $W_{N/S}$ ratio of **a** 4 and **b** 7. The variation of contact angle at different $W_{N/S}$ ratio is shown in (**c**)

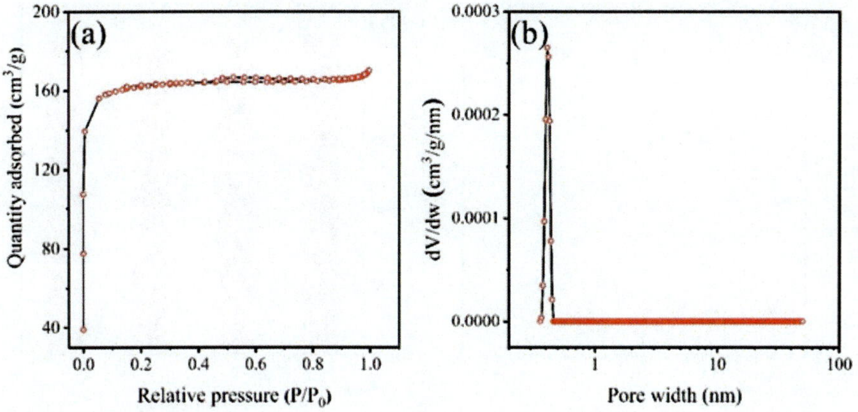

Fig. 35.5 **a** Nitrogen adsorption–desorption isotherms and **b** NLDFT pore-size distribution of carbon-foam sample ($W_{N/S} = 7$)

To check the absorption capacity and pore size present in the carbon-foam, the pore size distribution and specific surface area were calculated of $W_{N/S} = 7$ ratio. Figure 35.5a shows the representative N_2 adsorption–desorption isotherm of a carbon-foam ($W_{N/S} = 7$). At low relative pressure, there is a sharp absorption of N_2, indicating the presence of micropores [7, 10]. A narrow hysteresis is also observed at higher relative pressures which suggest the presence of mesopores [7, 10]. Using BET equation, the total specific surface area was calculated to be 537 m^2/g and the micropore area was calculated to be 456 m^2/g using the t-plot approach. The total micropore volume and pore volume of the carbon-foam were 0.2 and 0.26 cm^3/g, respectively. Figure 35.5b displays the carbon-foam pore-size distribution determined by the non-local density functional theory (NLDFT). The NLDFT graph indicates a peak corresponding to 0.4 nm.

To investigate the absorption performance, absorption capacities of samples of the carbon-foams developed at different $W_{N/S}$ for paraffin oil were measured. The weight of the solvent absorbed divided by the weight of the carbon-foam is known as the absorption capacity. It was noted that the absorption capacity increased with increasing the $W_{N/S}$ ratio up to 7. The carbon-foam with $W_{N/S} = 7$ shows a maximum absorption capacity for paraffin oil of 18.0267 g/g. After that the absorption capacity start decreasing (as shown in Fig. 35.6a). Furthermore, a high performance absorbent should also be reasonably recyclable for use in real-world applications in addition to being able to effectively separate organic liquids from water. Figure 35.6a depicts the absorption performance for paraffin oil of carbon-foam for different $W_{N/S}$ ratios.

Furthermore, to check the recyclability of the carbon-foam, the cyclic operations on a sample with $W_{N/S} = 7$ ratio was performed. The results indicated that the material ($W_{N/S} = 7$) has a stable absorption ability for paraffin oil during the cyclic absorption/evaporation activities. The recyclability of carbon-foam for paraffin oil is shown in Fig. 35.6b.

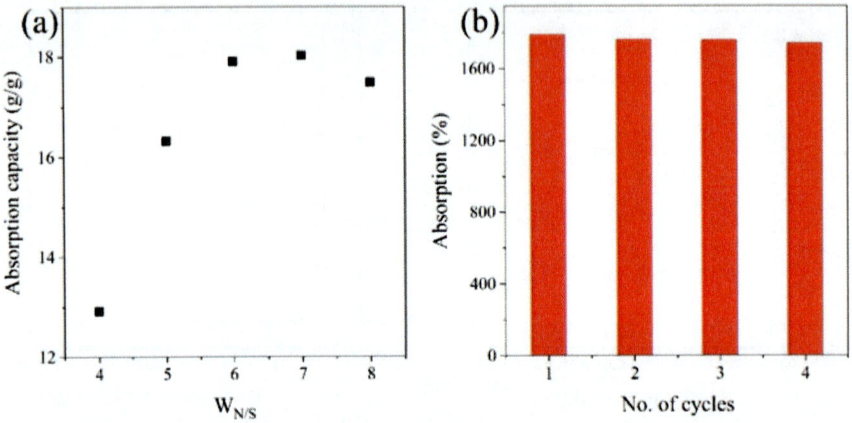

Fig. 35.6 **a** Absorption capacities of different $W_{N/S}$ ratio carbon-foams for paraffin oil and **b** recyclability of carbon-foam for paraffin oil

The carbon-foams produced from sucrose absorb oil phase from the water + paraffin oil mixture, as shown in Fig. 35.7. The photomicrograph of Fig. 35.7a depicts the 100 ml mixture of water and paraffin oil in the ratio of 90:10. The complete separation of paraffin oil from water + paraffin oil mixture took nearly 30 min upon adding the 5 g of carbon-foam from the $W_{N/S} = 7$ ratio sample (Fig. 35.7c). The carbon-foams employed for oil absorption from water–oil mixture could be regenerated by heating at 400 °C in N_2 atmosphere. The oil absorption ability remains almost steady even after 4 absorptions–regeneration cycles, as shown in Fig. 35.6b.

Fig. 35.7 **a** Photograph of the water + paraffin oil mixture and carbon-foam at 0 min, **b** partial paraffin oil absorption after 15 min and **c** complete paraffin oil absorption after 30 min

35.4 Conclusion

A simple method (compression moulding process) was employed for the synthesis of microcellular carbon-foam from sucrose, a natural renewable material along with NaCl particles as a low cost water soluble pore template. Extraction of NaCl particles using soxhlet extraction setup develops highly porous carbon-foam structure which is confirmed by the magnified SEM images. The density of the carbon-foams found decrease with increase in the $W_{N/S}$ ratio, which is due to the higher amount of NaCl. The density of the carbon-foam with $W_{N/S} = 4$ was found to be 0.0557 g/cm^3 and for $W_{N/S} = 8$ was 0.0332 g/cm^3. From the N_2 absorption measurements, the BET surface area was found to be 537 m^2/g and micropore area calculated by t-plot method was 456 m^2/g. The highest contact angle of 147° was observed for carbon-foam with $W_{N/S} = 7$ and there was no dependence of contact angle on the amount of NaCl addition. The carbon-foam with $W_{N/S} = 7$ shows a maximum absorption capacity for paraffin oil of 18.0267 g/g and excellent recyclability towards oil adsorption. The carbon-foam took around 30 min for complete separation of paraffin oil from water + paraffin oil mixture by adding the 5 g of carbon-foam from the $W_{N/S} = 7$ ratio sample.

Acknowledgements The authors sincerely thank the Department of Chemistry, Indian Institute of Space Science and Technology Thiruvananthapuram, Kerala, India, to provide the resources for conducting the current study.

Data Availability The raw/processed data required to reproduce these findings cannot be shared at this time as the data also forms part of an ongoing study.

References

1. Campagna, C., Short, F.T., Polidoro, B.A., Mcmanus, R., Collette, B.B., Pilcher, N.J., De Mitcheson, Y.S., Stuart, S.N., Carpenter, K.E.: Gulf of Mexico oil blowout increases risks to globally threatened species **61**, 393–397 (2011). https://doi.org/10.1525/bio.2011.61.5.8
2. Silva, I.A., Almeida, F.C.G., Souza, T.C., Sarubbo, L.A.: Oil spills: impacts and perspectives of treatment technologies with focus on the use of green surfactants. Environ. Monit. Assess. (2022). https://doi.org/10.1007/s10661-022-09813-z
3. Chau, M.Q., Truong, T.T., Hoang, A.T., Le, T.H.: Environmental effects oil spill cleanup by raw cellulose-based absorbents: a green and sustainable approach, energy sources, Part A recover. Util. Environ. Eff. **00**, 1–14 (2021). https://doi.org/10.1080/15567036.2021.1928798
4. Saleem, S., Hu, G., Li, J., Hewage, K., Sadiq, R.: Evaluation of offshore oil spill response waste management strategies: a lifecycle assessment-based framework. J. Hazard. Mater. **432**, 128659 (2022). https://doi.org/10.1016/j.jhazmat.2022.128659
5. Tuan, A., Ni, S.: Chemosphere advanced super-hydrophobic polymer-based porous absorbents for the treatment of oil-polluted water. **277** (2021). https://doi.org/10.1016/j.chemosphere.2021.130274
6. Vannarath, A., Thalla, A.K.: Synthesis and characterisation of an ultra-light, hydrophobic and flame-retardant robust lignin-carbon foam for oil-water separation. J. Clean. Prod. **325**, 129263 (2021). https://doi.org/10.1016/j.jclepro.2021.129263

7. Narasimman, R., Vijayan, S., Prabhakaran, K.: Carbon particle induced foaming of molten sucrose for the preparation of carbon foams. Mater. Sci. Eng. B Solid-State Mater. Adv. Technol. **189**, 82–89 (2014). https://doi.org/10.1016/j.mseb.2014.08.007
8. Narasimman, R., Vijayan, S., Prabhakaran, K.: Carbon foam with microporous cell wall and strut for CO_2 capture. RSC Adv. **4**, 578–582 (2014). https://doi.org/10.1039/c3ra46240d
9. Narasimman, R., Vijayan, S., Dijith, K.S., Surendran, K.P., Prabhakaran, K.: Carbon composite foams with improved strength and electromagnetic absorption from sucrose and multi-walled carbon nanotube. Mater. Chem. Phys. **181**, 538–548 (2016). https://doi.org/10.1016/j.matchemphys.2016.06.091
10. Vazhayal, L., Wilson, P., Prabhakaran, K.: Waste to wealth: lightweight, mechanically strong and conductive carbon aerogels from waste tissue paper for electromagnetic shielding and CO_2 adsorption. Chem. Eng. J. **381**, 122628 (2020). https://doi.org/10.1016/j.cej.2019.122628
11. Chithra, A., Wilson, P., Vijayan, S., Rajeev, R., Prabhakaran, K.: Carbon foams with low thermal conductivity and high EMI shielding effectiveness from sawdust. Ind. Crops Prod. **145**, 112076 (2020). https://doi.org/10.1016/j.indcrop.2019.112076
12. Liu, S., Huang, Z., Wang, R.: A carbon foam with a bimodal micro-mesoporous structure prepared from larch sawdust for the gas-phase toluene adsorption. Mater. Res. Bull. **48**, 2437–2441 (2013). https://doi.org/10.1016/j.materresbull.2013.02.069
13. Wang, M., Wang, C.Y., Li, T.Q., Hu, Z.J.: Preparation of mesophase-pitch-based carbon foams at low pressures. Carbon N. Y. **46**, 84–91 (2008). https://doi.org/10.1016/j.carbon.2007.10.038
14. Lei, S., Guo, Q., Shi, J., Liu, L.: Preparation of phenolic-based carbon foam with controllable pore structure and high compressive strength. Carbon N. Y. **48**, 2644–2646 (2010). https://doi.org/10.1016/j.carbon.2010.03.017
15. Prabhakaran, K., Singh, P.K., Gokhale, N.M., Sharma, S.C.: Processing of sucrose to low density carbon foams. J. Mater. Sci. **42**, 3894–3900 (2007). https://doi.org/10.1007/s10853-006-0481-1
16. Moglie, F., Micheli, D., Laurenzi, S., Marchetti, M., Mariani Primiani, V.: Electromagnetic shielding performance of carbon foams. Carbon N. Y. **50**, 1972–1980 (2012). https://doi.org/10.1016/j.carbon.2011.12.053
17. Szczurek, A., Fierro, V., Pizzi, A., Stauber, M., Celzard, A.: A new method for preparing tannin-based foams. Ind. Crops Prod. **54**, 40–53 (2014). https://doi.org/10.1016/j.indcrop.2014.01.012
18. Narasimman, R., Prabhakaran, K.: Preparation of carbon foams with enhanced oxidation resistance by foaming molten sucrose using a boric acid blowing agent. Carbon N. Y. **55**, 305–312 (2013). https://doi.org/10.1016/j.carbon.2012.12.068

Chapter 36
Microwave Heating Mechanism of AZ31/HA Metal Matrix Biocomposites

Shivani Gupta, Apurbba Kumar Sharma, Dinesh Agrawal, and Inderdeep Singh

Abstract In the present scenario, advancements in biocomposite processing have become the prime research domain for technocrats and manufacturers. Understanding the processing technique is a challenging task: what is the mechanism, how does the selected process work and influence the material properties and how does it control the acceptance/rejection of the end product; these activities are enclosed. This work focuses on understanding the microwave biomaterial processing mechanism and the effects of microwaves on the processed biocomposite features. Microwave sintering process is used to process AZ31/HA biocomposites. Microwave heating mechanism of HA (bioceramic with dipolar losses) and AZ31 metal alloy (conduction loss), with the help of schematic and suitable explanations, is also analyzed. In addition, the effects of microwave radiation on the developed AZ31/HA biocomposites are investigated. It occurred due to the addition of HA powder particles that improved the heating rate and created uniform heating inside the composite volume, behaving like a localized heating source for the neighboring AZ31 powder particles inside the composite volume.

Keywords Electromagnetic energy · Microwaves · Powder metallurgy · Biocomposites · Mechanical properties

S. Gupta (✉) · A. K. Sharma · I. Singh
Mechanical and Industrial Engineering, Indian Institute of Technology Roorkee, Roorkee, India
e-mail: sgupta2@me.iitr.ac.in; shivani.gupta@mitwpu.edu.in

S. Gupta
MIT World Peace University, Pune, India

D. Agrawal
Materials Research Institute, Pennsylvania State University, Pennsylvania, USA

A. Tewari et al. (eds.), *Proceedings of the 3rd International Conference on Advances in Materials Processing: Challenges and Opportunities*, Springer Proceedings in Physics 293, https://doi.org/10.1007/978-981-99-1971-0_36

36.1 Introduction

Selection of the suitable manufacturing process for the biomaterials (BMs) has become a promising task for researchers as manufacturing technique plays a prime role in the end product's performance. It controls material performance by modifying original properties. Thus, the modification in material properties can be observed by understanding the mechanism of the processing technique. It includes conventional/nontraditional machining, additive manufacturing, casting, and powder metallurgy routes in which many advancements have already been accomplished and support new energy sources and accessories [1]. Similarly, in the powder metallurgy route, researchers are trying to find the desired outcomes by altering the energy sources and processing techniques such as liquid phase sintering, spark plasma sintering, flash sintering, microwave sintering, etc. [2, 3]. Furthermore, many people have worked on microwave processing of BMs and achieved improved properties compared to conventional ones [4, 5]. Literature is available on hydroxyapatite (HA) powder production through microwave synthesis and further used as artificial BMs [6–8]. People also sintered HA-based composites using microwave energy and analyzed the process through modeling and simulation tools to understand the process's physics [9–12]. Mg/HA composites were also developed through microwave sintering and achieved good mechanical and biological properties [13, 14]. Many studies have been conducted on the microwave material interaction phenomenon and mechanism [15]. However, significantly less literature is available to understand the exact phenomenon and mechanism of microwave processing of various BMs. Therefore, this work is attempted to explore how microwaves interact with composites. AZ31-HA metal matrix biocomposites were developed through the microwave sintering process and explored the microwave heating and sintering mechanism.

36.2 Materials and Methods

Magnesium-aluminum-zinc-based alloys (AZ31) and HA were used as matrix and reinforcement, respectively. Both powders were mixed with the help of a V-type blender in various compositions. The powder mixture was pressed at 350 MPa hydraulic pressure in a cylindrical shape of 13 mm diameter and 5 mm thick. Prepared pellets were sintered using a multimode microwave applicator (Make: Enerzi Microwave Systems Pvt. Ltd. Belgaum, Karnataka, India), operating at 2.45 GHz frequency and 1.4 kW power. The samples were sintered at 500 °C temperature. After that, densification and microstructures were measured and the effect of microwave energy on the properties of microwave-sintered biocomposites was analyzed.

36.3 Results and Discussion

This section discusses the microwave heating mechanism and microwave interaction with AZ31 and HA powder particles. In addition, the microwave sintering mechanism is also described with the help of suitable schematics and explanations. Characteristics of developed biocomposites are also discussed.

36.3.1 Microwave-Material Interaction

Microwaves are electromagnetic waves having a frequency range from 300 MHz to 300 GHz and wavelength from 1 mm to 1 m. These are the combination of electric and magnetic fields [16]. The electrical and magnetic properties of the material decide the microwave energy absorption phenomenon in microwave materials processing. Maxwell's theories and their equations govern the changes in the electric and magnetic fields with time domain and material properties [17]. Complex permittivity is the prime factor in measuring heat generation in a non-magnetic material. Microwave absorption is considered through 'Penetration Depth' and 'Skin Depth'. The former is the primary characteristic for non-metallic and later for metallic materials [15]. Absorbed microwave power depends on the material's electromagnetic attributes, shape, and size. The microwave penetration and absorbed power are calculated from Eqs. 36.1–36.3 [16].

$$D_p = \frac{1}{\sqrt[\omega]{\left[0.5\mu_0\mu'\varepsilon_0\varepsilon'\left\{\sqrt{1+\left(\frac{\varepsilon''_{eff}}{\varepsilon'}\right)^2}-1\right\}\right]}} \tag{36.1}$$

$$D_s = \frac{1}{\sqrt{(\pi f \mu\sigma)}} = 0.029(\rho\lambda_0)^{0.5} \tag{36.2}$$

$$P = 2\pi f \varepsilon_0 \varepsilon'' E_{rms}^2 \tag{36.3}$$

$$\sum mC\Delta T = 2\pi f \varepsilon_0 \varepsilon'' E_{rms}^2 V t \tag{36.4}$$

where ε_0 is the permittivity of the material in air (Fm^{-1}), ε''complex permittivity (Fm^{-1}), μ magnetic permeability (Hm^{-1}), μ_0 magnetic permeability of material in air (Hm^{-1}), μ' imaginary part of magnetic permeability (Hm^{-1}), f frequency of microwave (s^{-1}), σ is electrical conductivity (Sm^{-1}), ρ electrical resistivity (Ω m) and λ_0 (m) wavelength of incidence radiation. *P* is absorbed microwave power per unit volume (Wm^{-3}), and E_{rms} is the root mean square of the electric field ($V\ m^{-1}$), m (kg) is the mass of the targeted material, C ($Jkg^{-1}\ K^{-1}$) is the specific heat of the targeted material, ΔT (°C) is the temperature difference, V (m^3) is the volume of

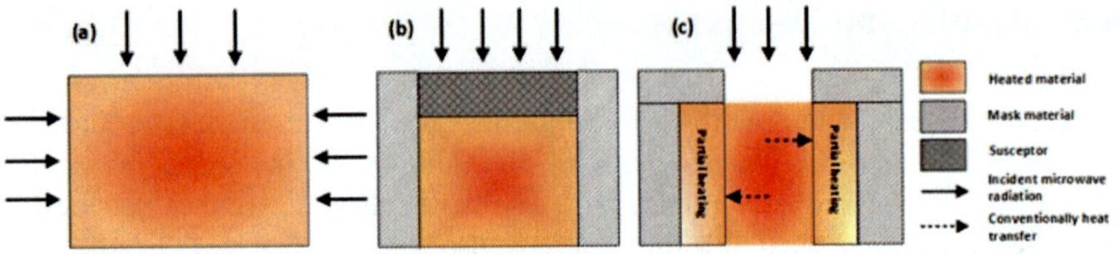

Fig. 36.1 Methods of microwave heating; **a** direct heating, **b** selecting heating, and **c** hybrid heating

the material and *t* (s) is the exposure time. The absorbed power makes the targeted material heat up and increases the temperature, which can be expressed in Eq. 36.4. In addition, the heating mechanism plays a significant role in microwave material interaction and processing. Microwave heating can be achieved through three types of methods such as (a) direct heating, (b) selective heating, and (c) hybrid heating, in which the susceptor and masking material are needed to accomplish effective material processing. Figure 36.1 shows the different heating techniques.

In addition, the dipolar loss is an efficient factor in microwave heating of insulators like ceramics, CMC, PMC, and water, in which dipoles are present and turn out to be energetic in the presence of the electric field. Similarly, in the case of metallic materials, conduction loss (due to the presence of free electrons) is the primary source of microwave heating [18].

36.3.2 Microwave Interaction with Targeted Materials

This study used the direct heating method to process the AZ31/HA biocomposites. AZ31 is a metal alloy, while hydroxyapatite (HA) is a bioceramic that is a good absorber of microwaves. AZ31 metal alloy exhibited reflective nature in exposure to microwaves due to the presence of free electrons. Free electrons generated their own current and magnetic field that acted opposite to the external electric field component resulting in the polarization of free electrons. The skin depth of AZ31 metal alloy was calculated from Eq. 36.2 and listed in Table 36.1. AZ31 alloy and HA are both non-magnetic materials. The chemical structure of HA is $Ca_{10}(PO_4)_6(OH)_2$, which reveals the presence of hydroxyl ions (OH^-), act as dipoles and assists in being a good absorber of microwaves. HA ceramic has a higher dielectric loss than AZ31 [18]. Dielectric characteristics, such as ε', ε'' and quality factor (Q) of cortical bone, AZ31 metal alloy, and HA are listed in Table 36.1. The heating mechanism in the development of biocomposites is conduction and dipolar losses due to the material characteristics [19]. The microwave material interaction mechanism in targeted materials; AZ31/HA biocomposites is shown in Fig. 36.2. The mechanism of microwave interaction in HA powder particles is shown in Fig. 36.2a. The microwave absorption occurred due to dipole movement (hydroxyl ions). As microwaves entered the HA particle, existing hydroxyl ions started rotating due to the external electric field (E) in the first half cycle and rotated in the opposite direction in the next half cycle. This motion of hydroxyl ions produced oscillation in the material and resulted in heat generation. This heat generation was because of energy conversion (microwave energy into heat) due to vibration, oscillation, and kinetic motion of ions. Similarly, Fig. 36.2b demonstrates the microwave interaction with non-magnetic AZ31 metal alloy. As microwave radiations entered the metal alloy particle, all existing free electrons started to move in the direction of the present electric field with velocity (v) and created an electron cloud in the first half cycle. In the next half cycle, these free electrons generated their induced magnetic field and current that worked in the opposite direction of the external electric field. Due to such induced current, these free electrons moved with relative velocity (v_r) and reached their original positions again. Such movement of free electrons generated oscillation, resulting in heat generation in the material and raising temperature [20]. Both materials behave differently in microwaves due to different electrical conductivity, dielectric constant, dielectric loss, and interaction mechanisms with microwaves.

Table 36.1 Microwave properties of natural bone and interested constituents of biocomposites

Material	Dielectric properties		Electrical resistivity	Penetration depth/skin depth	Reference
	ε'	ε''	ρ ($\times 10^{-8}$ Ω m)	D_P	
Cortical bone at 2.45 GHz	11.38	2.89	–	2.29 cm	Sharma and Gupta [22]
HA at 100 kHz and 400 K	20.4	8.2	–	–	Ikoma et al. [18]
AZ31 at 2.45 GHz			4.13	2.1 μm	Equation 36.2

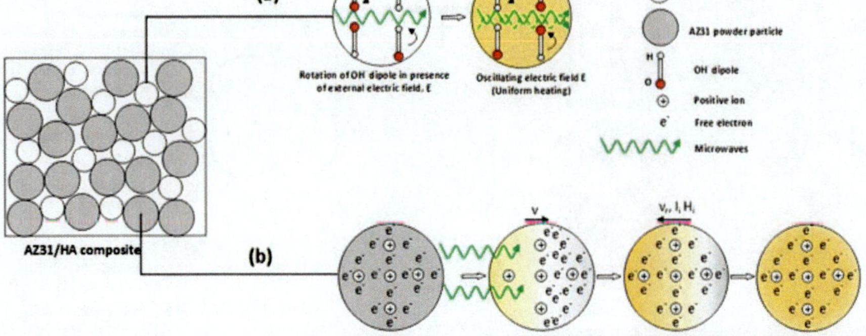

Fig. 36.2 Microwaves material interaction in targeted materials: **a** HA and **b** AZ31 powder particles

36.3.3 Microwave Sintering Mechanism

Microwave energy is used for heating the targeted biocomposites. The biocomposites were heated up to their respective sintering temperature. The microwave radiations penetrated the green pellet and generated heat for powder diffusion and necking formation between the powder particles [21]. Figure 36.3 depicts the microwave heating mechanism with different weight percentages of HA in biocomposites. HA attributes a high dielectric loss factor that enhances the heating rate and results in isothermal heating [22]. HA particles absorb more microwave energy than AZ31 and behave like a tiny heat source for the surrounding metal alloy particles. These localized heat sources transfer the heat through conduction mode to the nearby metal alloy particle, resulting in an increased temperature of the pellets. Thus, these HA particles increase the uniform heating effect throughout the pellet volume. In Fig. 36.3a a small amount of HA particles is present, which transfers less amount of heat to the metal particles. On the other hand, in Fig. 36.3b, HA particles are present in large amounts, generate more heat and transfer it uniformly to the neighboring particles.

In addition, the same pattern was observed during microwave heating of biocomposites. As HA content increased, the heating rate also increased and resulting in a short processing time. The targeted sintering temperature of 500 °C was achieved in a short time in the case of composites as compared to the AZ31-0HA sample. Further, the density of developed biocomposites was measured and found that a high HA percentage composition depicts high density due to more uniform heating. They also depicted high microhardness due to the presence of hard ceramic phases like HA. The findings of the study are listed in Table 36.2. The density of biocomposites

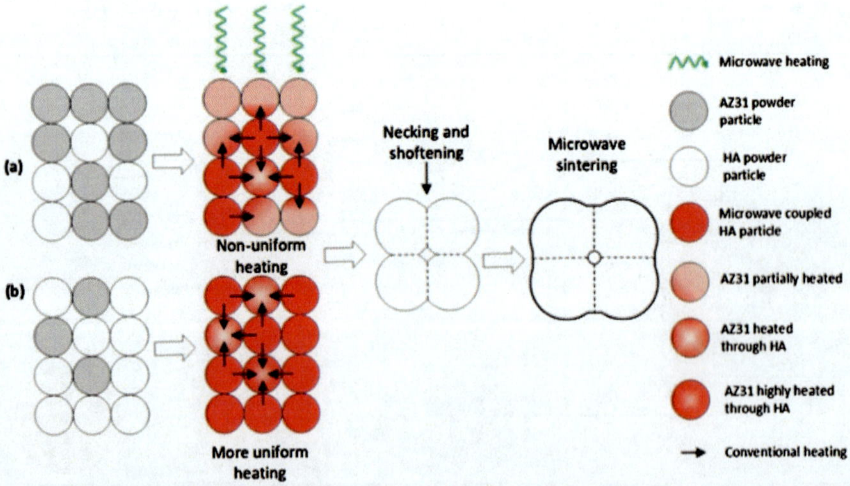

Fig. 36.3 Schematic of microwave sintering mechanism happened in **a** a small percentage of HA and **b** a large percentage of HA powder particles in AZ31/HA biocomposites [23]

increased with increasing HA content and a similar pattern was observed in microhardness. AZ31-15HA depicts the highest density of 97.5% and 76 HV microhardness. Table 36.2 also depicts the processing time to attain 500 °C in various compositions. The experimental results correlate with microwave materials interaction and heating mechanism (as shown in Figs. 36.2 and 36.3). The recorded processing time shows that the addition of HA content reduces the processing time. Figure 36.4 shows the SEM microstructures of AZ31-0HA, AZ31-10HA, and AZ31-15HA samples along with their EDX elemental analysis. SEM microstructures represent the presence of Mg and MgO over the sample of pure AZ31, while in biocomposites, HA phase is observed. HA particles are distributed at the boundaries of Mg particles. HA particles of nano-size accumulated at the surface. The presence of HA in biocomposites improved the density and microhardness, and it can be confirmed by the microstructures also. Figure 36.4e shows a higher dense structure than Fig. 36.4a and c.

Table 36.2 Density, porosity, and microhardness of microwave sintered biocomposites

Sample	Processing temperature (°C)	Processing time (min)	Density (g/cm^3)	Relative density (%)	Porosity (%)	Microhardness (HV)
AZ31–0HA	500	20 ± 2	1.68	94.5	5.5	60
AZ31–10HA	500	15 ± 1.5	1.82	96.2	3.8	71
AZ31–15HA	500	12 ± 2	1.88	97.5	2.5	76

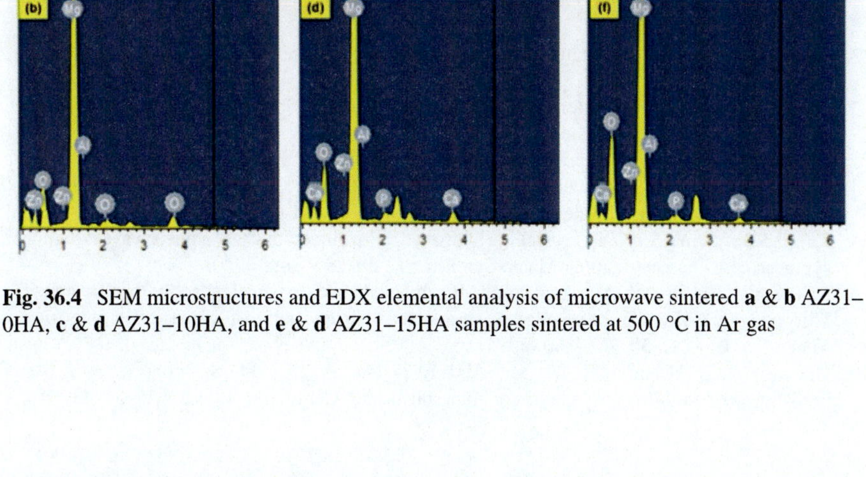

Fig. 36.4 SEM microstructures and EDX elemental analysis of microwave sintered **a** & **b** AZ31–0HA, **c** & **d** AZ31–10HA, and **e** & **d** AZ31–15HA samples sintered at 500 °C in Ar gas

36.4 Conclusion

In this article, the microwave heating mechanism is studied through appropriate electromagnetic theories and schematic diagrams of microwave interaction with individual constituents of the biocomposites at their atomic level. HA is bioceramic and has high dielectric loss than AZ31 metal alloy. It absorbs more microwave radiation because of OH^- (hydroxyl) ions that provide dipolar moment and heat the material rapidly. The addition of HA powder particles helped to enhance the heating rate with uniform heating during microwave materials interaction. On the other hand, AZ31 metal alloy has free electrons, reflects microwaves, and contributes less to the microwave heating of biocomposites. Additionally, microwave sintering mechanism in AZ31/HA biocomposites is also elaborated with the help of suitable diagrams and explanations. It shows good necking and diffusion during microwave sintering due to volumetric heating. It happened due to more absorption of microwaves and provided more, and uniform heat generation. The reverse heating phenomenon in microwave heating (gets heat up from the core to the surface) also reduced the processing time along with improved density, microhardness, and microstructures. It is concluded that the microwave heating and sintering mechanisms explained in this article can be used in understanding microwave interaction with various biomaterials processing in the future.

Acknowledgements This work is financially supported by the Ministry of Education, Government of India under SPARC scheme through its Project No. SPA-1422-MID. The authors acknowledge the infrastructural and other support received from Indian Institute of Technology Roorkee and Pennsylvania State University, USA, for the collaboration work.

References

1. Gupta, S., Sharma, A. K.: Sintering of biomaterials for arthroplasty: a comparative study of microwave and conventional sintering techniques. Appl. Mech. Mater. **895**. Trans Tech Publications Ltd, (2019).
2. Kang, C.-W., Fang, F.-Z.: State of the art of bioimplants manufacturing: part I. Adv. Manuf. **6**(1), 20–40 (2018)
3. Eriksson, M., Liu, Yi., Jiangfeng, Hu., Gao, L., Nygren, M., Shen, Z.: Transparent hydroxyapatite ceramics with nanograin structure prepared by high pressure spark plasma sintering at the minimized sintering temperature. J. Eur. Ceram. Soc. **31**(9), 1533–1540 (2011)
4. Lerner, E., Sarig, S.: Enhanced maturation of hydroxyapatite from aqueous solutions using microwave irradiation. J. Mater. Sci.: Mater. Med. **2**, 138–141 (1991)
5. Kalita, S.J., Verma, S.: Nano-crystalline hydroxyapatite bioceramic using microwave radiation: synthesis and characterization. Mater. Sci. Eng.: C. **30**, 295–303 (2010)
6. Radha, G., Balakumar, S., Venkatesan, B., Vellaichamy, E.: Evaluation of hemocompatibility and in vitro immersion on microwave-assisted hydroxyapatite–alumina nano-composites. Mater. Sci. Eng.: C. **50**, 143–150 (2015)
7. Hassan, M. N., Mahmoud, M. M., Abd El-Fattah, A., Kandil, S.: Microwave-assisted preparation of nano-hydroxyapatite for bone substitutes. Ceram. Int. **42**(3), 3725–3744 (2016).

8. Shi, X., Zhou, J., Liu, G., Wang, L.: The physical and antimicrobial properties of silver doped hydroxyapatite sintered by microwave and conventional sintering. J. Inorg. Organomet. Polym Mater. **27**(4), 955–961 (2017)
9. Choy, M. T., Tang, C. Y., Chen, L., Wong, C. T., Tsui, C. P.: In vitro and in vivo performance of bioactive Ti6Al4V/TiC/HA implants fabricated by a rapid microwave sintering technique. Mater. Sci. Eng.: C. **42**, 746–756 (2014).
10. Birnboim, A., Carmel, Y.: Simulation of microwave sintering of ceramic bodies with complex geometry. J. Am. Ceram. Soc. **82**(11), 3024–3030 (1999)
11. Riedel, H., Svoboda, J.: Simulation of microwave sintering with advanced sintering models. In: Advances in Microwave and Radio Frequency Processing, pp. 210–216. Springer, Berlin, Heidelberg (2006).
12. Xiong, G., Nie, Y., Ji, D., Li, J., Li, C., Li, W., Zhu, Y., Luo, H., Wan, Y.: Characterization of biomedical hydroxyapatite/magnesium composites prepared by powder metallurgy assisted with microwave sintering. Curr. Appl. Phys. **16**(8), 830–836 (2016).
13. Sunil, B.R., Ganapathy, C., Kumar, T.S., Chakkingal, U.: Processing and mechanical behavior of lamellar structured degradable magnesium hydroxyapatite implants. J Mech Behav Biomed Mater **40**, 178–189 (2014)
14. Zhang, H., Datta, A.K.: Microwave power absorption in single-and multiple-item foods. Food Bioprod. Process. **81**(3), 257–265 (2003)
15. Mishra, R. R., Sharma, A. K.: Microwave–material interaction phenomena: heating mechanisms, challenges and opportunities in material processing. Compos. Part A: Appl. Sci. Manuf. **81**, 78–97 (2016).
16. Mishra, R. R., Sharma, A. K.: A review of research trends in microwave processing of metal-based materials and opportunities in microwave metal casting. Crit. Rev. Solid State Mater. Sci. **41**(3), 217–255 (2017).
17. Gupta, M., Leong, E. W. W.: Microwaves and metals. John Wiley & Sons (2008).
18. Ikoma, T., Yamazaki, A., Nakamura, S., Akao, M.: Preparation and dielectric property of sintered monoclinic hydroxyapatite. J. Mater. Sci. Lett. **18**(15), 1225–1228 (1999)
19. Sharma, A. K., Mishra, R. R.: Role of particle size in microwave processing of metallic material systems. Mater. Sci. Technol. **34**(2), 123–137 (2018).
20. Roy, R., Agrawal, D., Cheng, J. P.: New first principles of microwave-material interaction: discovering the role of the H field and anisothermal reactions. In: Clark, D. E., Jon G. P., Binner, Lewis, D. A. (eds.) Microwaves: Theory and Application in Materials Processing V, vol. 111, pp. 471–485. Ceramic Trans, ACS Publ (2001).
21. Singh, S., Gupta, D., Jain, V., Sharma, A.K.: Microwave processing of materials and applications in manufacturing industries: a review. Mater. Manuf. Processes **30**(1), 1–29 (2015)
22. Sharma, A. K., Gupta, S.: Microwave processing of biomaterials for orthopedic implants: challenges and possibilities. JOM. **72**(3), 1211–1228 (2020).
23. Gupta, S., Sharma, A.K., Agrawal, D., Lanagan, M.T., Sikora, E., Singh, I.: Characterization of AZ31/HA biodegradable metal matrix composites manufactured by rapid microwave sintering. Materials. **16**(5), 1905 (2023).

Chapter 37
Recovery of Tungsten and Copper from Tungsten-Copper (W-Cu) Electrical Contacts Borings

Manoj Kumar and Maitreyee Bhattacharya

Abstract A leaching-solvent extraction-electrowinning-reduction process is developed to recover the copper and tungsten from tungsten-copper (W-Cu) borings, generated during the production of tungsten-copper (W-Cu) composite electrical contacts. Direct ammonia leaching was applied to dissolve the W-Cu turnings of ground size –20, +140 mesh. Dissolution of upto 95.0% was achieved at the optimized condition of 125 g/L ammonia concentration, 1500 rpm stirring speed, 125 g/L solid:liquid ratio in 7h time at 500g scale. The leach liquor containing 26.58 g/L W, 16.03 g/L Cu and 85.56 g/L ammonia was used for solvent extraction of copper using 25% LIX 84 (2-hydroxy-5-nonylacetophenone oxime) in an appropriate organic to liquid ratio. The Cu loaded organic LIX 84 was stripped using 20 vol. % sulphuric acid. Copper as copper powder was recovered from the stripped solution containing 12.70 g/L copper used for electrowinning. Current efficiency of 38.95% was found at cathode current density of 725 A/m^2. The recovery of tungsten from the separated tungsten containing ammoniacal solution was done in a Rotary Vacuum Evaporator as ammonium paratungstate (APT). APT was reduced to high purity tungsten powder of 99.8% in a tubular furnace. At optimum process parameters of 1000 °C temperature, 0.1 LPM hydrogen flow rate in 3h reduction was 99.99%. Increasing the bed depth was found to decrease the reduction.

Keywords Tungsten-Copper (W-Cu) borings · Tungsten powder · Copper powder

37.1 Introduction

Tungsten is a unique and strategic metal due to high strength at high temperature. It has wide applications in industrial sectors, such as energy, electrical contact composite materials, information technology, metalworking, mining and

M. Kumar (✉) · M. Bhattacharya
Metal Extraction and Recycling Division, CSIR-National Metallurgical Laboratory(CSIR-NML), Jamshedpur 831007, India
e-mail: mk@nmlindia.org

A. Tewari et al. (eds.), *Proceedings of the 3rd International Conference on Advances in Materials Processing: Challenges and Opportunities*, Springer Proceedings in Physics 293, https://doi.org/10.1007/978-981-99-1971-0_37

stone-cutting tools and aerospace [1]. According to the 2021 Mineral Commodity Summaries on tungsten by the US Geological Survey (USGS), the major tungsten-producing countries around the world are China, Vietnam, Russia, Mongolia, Bolivia, Austria, and Rwanda [2]. In 2020, the global tungsten output was about 84 kt. Among major tungsten-producing countries, China remained the largest supplier, and accounts for more than 82% of the world's production, reaching about 69 kt [2]. A scarce of potential substitutes and geographically concentrated production make tungsten a critical element in the global economy [3]. Currently, primary mining covers most of the demand for this metal, with recycling mainly limited to industrial scrap.

The recovery of tungsten from tungsten bearing scrap generated in various industries is an important source of valuables. It has already been used as a substitute for tungsten ore concentrates at conversion plants [4]. Tungsten scrap *viz.* metal drill scrap, super alloys, and tungsten carbide products scrap, and powder metallurgical tungsten new scrap could be divided into three types: new scrap (by-products of tungsten materials or products), old scrap (spent tungsten-bearing materials), and unrecovered scrap (excess reactants) [1]. The tungsten content in tungsten bearing scrap is in the range of 40–95 wt. % [4]. Since 2000, a wide variety of tungsten scrap recycling methods have been used globally. Approximately 30% of world tungsten demand is met through tungsten scrap recycling [5].

Different methods *viz.* zinc process, cold stream process, oxidation-reduction process, electrolysis process etc. are employed for the processing of secondary tungsten scrap generated from various resources [6–8]. Varieties of tungsten containing alloy scraps were treated by various processes depending upon the values and impurities present. The conventional processes in which both soft and hard scraps are usually roasted in the presence of alkali salts of sodium carbonate at 650−750 °C and then they were leached in sodium hydroxide to dissolve tungsten as sodium tungstate [9]. The leached tungsten values and the impurities are separated and recovered by the various hydrometallurgical processes. Ammonium paratungstate (APT) is produced from the sodium tungstate solution resulting from the processing through the several steps [10]. Though the sodium nitrate/sodium carbonate fusion process is employed for recovering the metallic values, generation of nitrogen oxides causes environmental pollution [11]. The tungsten of soft alloy scrap (turnings of W-Cu alloy) was also electrochemically dissolved to the extent of 90% in sodium hydroxide using packed bed of scrap, apart from other processes to obtain the very low level of impurities contaminated during the processing [12].

The production of tungsten metal powder is carried out almost exclusively by the hydrogen reduction at temperatures in the range of 900–1000 °C of high-purity tungsten oxides (WO_3, WO_{3-x}) present of intermediate product ammonium paratungstate (APT), obtained during processing of either extraction of tungsten from ores or recovery from tungsten-scrap [13–15]. This paper contributes to the development of the tungsten and copper recovery as pure tungsten and copper powder from the tungsten-copper (W-Cu) composite material, generated during the production of electrical contacts by a novel and sustainable process of leaching-solvent extraction-electrowinning-reduction in an environmentally friendly and economical way.

37.2 Theoretical Consideration

Ammonia leaching is based on its ability of aqueous ammonia to form amine complexes which stabilize metals in solution at alkaline pH. Ammonia can exist in several forms, and these include free ammonia (NH_3), ammonium ions (NH4+) and metal ammines. In the Cu-NH_3-H_2O system, the species present are NH_3, NH4+, H+, Cu^{2+}, Cu^{+}, $Cu(OH)^{3-}$, $Cu(OH)_4{}^{2-}$, $Cu(OH)_2{}^{2-}$, $CuNH^{3+}$, $Cu(NH_3)_2{}^{2+}$, $Cu(NH_3)^{2+}$, $Cu(NH_3)_2{}^{2+}$, and $Cu(NH_3)_4{}^{2+}$. The stability of these complexes is dependent on the pH and Eh values [16]. In ammoniacal leaching of W-Cu boring, both copper and tungsten highly dissolve simultaneously at pH >10 and form tungsten-copper-ammine complex. Therefore, copper dissolves as per reaction:

$$Cu^{2+} + NH_4OH = Cu\ (NH_3)_4^{2+} + H_2O$$

In ammoniacal leaching of tungsten-copper, tungsten may dissolve in ammonia solution forming the tungsten-copper-amine complex. The reaction of dissolving tungsten into ammonia solution may be represented by

$$W + NH_4OH = (NH_4)_2WO_4 + H_2O$$

Ammonium tungstate solution left over, after separation of copper by solvent extraction using LIX 84, was then slowly evaporated at temperature above 80 °C in a rotary vacuum evaporator under vacuum in the range of 400–500mm of Hg for crystallization. The crystallized ammonium para tungstate (APT) $(NH_4)_{10}[H_2W_{12}O_{42}].4H_2O$ was reduced by hydrogen gas. The details of the reactions taking place during reduction are reported by Kumar et al. [17].

37.3 Experimental

37.3.1 Materials

Tungsten-copper borings (W-Cu borings) obtained from the M/s Mipalloy Chennai containing 53.78% Cu, 0.13% Fe, 46.01% W and 0.0014 Co was used in this investigation. The original supplied size of 4 to 15 mm W-Cu borings turnings were ground to the –20 + 140 mesh size and used for leaching.

37.3.2 Procedure

The W-Cu borings was cleaned and washed in water. It was used for experiments at 500g scale in a special reactor of capacities 5L. The amounts of W-Cu alloy turnings, dried in oven for more than 2h, were taken as per the desired solid-liquid ratios upto 500g scale. The same was stirred using stainless steel stirrer having PTFE lining at 500g scale at room temperature (25 °C). The stirring speed varied from 500 to 2000 rpm. After running the experiments under different leaching conditions, the leach solutions were filtered using the vacuum filtration machine. The residue left at the bottom of the reactor after leaching was dried in an oven at 110 °C for more than 2 hrs. Weight of the dried sample was noted down. Accordingly, the percentage of dissolution of tungsten and copper was calculated. The filtered leach solutions were chemically analyzed for the dissolved W and Cu by AAS and ICP.

% Leaching = $W_o - W_1 / W_o$
W_o—Initial weight of the sample
W_1–Final weight of the sample (dried residue left after filtration)

The 25% LIX84 (2-hydroxy-5-nonylacetophenone oxime) was used as the organic solvent to separate the copper from the complex solution containing tungsten and copper. The aqueous solution was analyzed for copper, tungsten, iron, cobalt etc. using AAS/ ICP. The raffinate solution obtained from the actual leach liquor was processed for the recovery of tungsten values by the crystallization process. For the leach liquor generated from tungsten-copper borings containing 26.58 g/L W, 16.03 g/L Cu and 85.56 g/L ammonia was used for the solvent extraction.

All experiments were performed in an electrochemical cell of capacity 1.5 L to recover copper. Unless otherwise stated the electrolytes used were 1 L of 40 g/L Cu in the copper sulfate solution, temperature was 25 °C and the current density maintained was 725 and 1250 A/m^2. In synthetic acid solution titanium plate was used as the anode and lead plate was used as the cathode, whereas in the actual stripped solution a stainless steel rod was used as the cathode and lead-antimony plate was used as the anode. The brush-down time interval of copper powder deposited was kept as 1h for the experiments for about 4–7 h duration. The obtained copper powder was dried and separated in different size fractions using sieve shaker were weighed were calculated for the various weight fractions against each size.

Ammonium paratungstate (APT) crystallized in Rotary Vacuum Evaporator APT, obtained from the above tungsten containing raffinate solution, was directly reduced in a tubular furnace powder to produce the tungsten powder using the hydrogen gas. The even temperature heating zone of 10 cm of the tubular furnace was utilized for reducing the ammonium paratungstate kept in a boat of size 8 cm. The effect of bed depth was investigated at certain optimized process parameters of temperature, hydrogen gas flow rate and time. Purity of the tungsten powder obtained was determined by ICP. Tungsten powder was also characterized by EPMA.

Table 37.1 Ammonia leaching of tungsten-copper electrical contacts borings

S.No.	W-Cu borings (g)	Solid: Liquid (g/L)	Time (h)	Leaching (%)
3	500	250	5	56.7
5	500	125	5	77.0
6	500	100	7	95.0

37.4 Results and Discussion

37.4.1 Leaching of Tungsten-Copper Borings at 500 g Scale

The ammonia leaching investigations of tungsten-copper turnings at 500 g scale in the 5 L reactor were carried out at different stirring speed, that is, 1500 rpm stirring speed with the variation in solid : liquid ratio of 250, 125, and 100 g/L at the ammonia concentration of 125 g/L for the time intervals of 5 and 7 h. The leaching percentage increases from 56.7 to 95.0 %. The results are given in Table 37.1. The leaching percentage of W-Cu turnings of –20 +140 mesh size increases upto 95% at the 100 g/L solid: liquid ratio, 1500 rpm stirring speed and 7 h time.

37.4.2 Electrowinning of Copper from the Synthetic and Actual Stripped Solution

The electrowinning experiments were carried out from the synthetic solution containing 40 g/L copper concentration in the copper sulfate, 160 g/L sulfuric acid concentration and for 5h duration. The brush down time interval was kept constant as 1h. The effect of current density on the current efficiency and the power consumption was shown in Fig. 37.1.

It is evident that with increase in the cathode current density from 600 to 800 A/m^2 the current efficiency increases from 45.2 to 94.9% and accordingly, the power consumption decreases from 9.88 kWh/kg copper powder to 3.38%. Beyond that the increase in the current density from 800 to 1400 A/m^2 the current efficiency decreases at a slow rate from 94.9 to 86.2%. This has led to an increase in the power consumption from 3.38 kWh/kg copper powder to 5.46 kWh/kg copper powder. This may be resulting from an increase in current density higher than 800A/m^2 increases the cathode polarization causing more hydrogen to be discharged.

Electrowinning of copper powder investigations were conducted from stripped solutions containing 12.70 and 4.32 g/L copper in a 36g/L sulfuric acid solution at

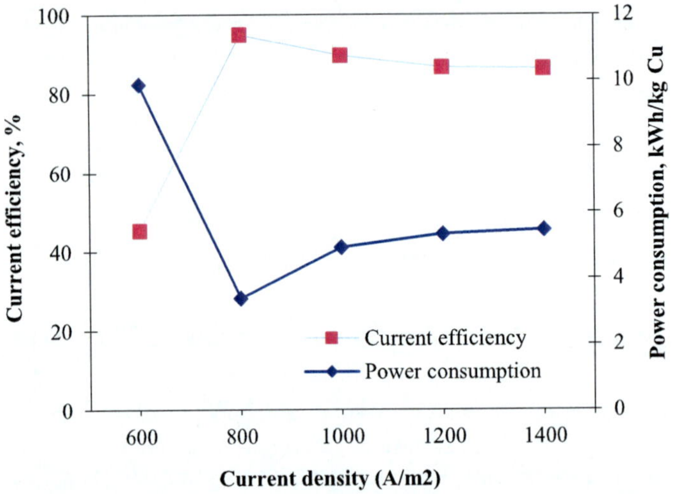

Fig. 37.1 Effect of current density on current efficiency and power consumption during electrowinning of copper powder

Table 37.2 Electrowinning from 1L stripped $CuSO_4$ solution

Cu in stripped solution (g/L)	Cathode current density (A/m^2)	Time (h)	Copper powder deposited (g)	Current efficiency (%)	Power consumption (kWh/kg)
4.32	1250	4	4.76	22.29	22.31
4.32	725	4	4.0	24.79	19.72
12.70	725	7	1 1.0	38.95	12.12

the cathode current density parameters of 725 and 1250A/m^2 for the time intervals of 4 and 7h (Table 37.2). With lowering of current density from 1250 to 725A/m^2 the current efficiency was 22.29% and 24.79%, respectively, in a 4 h duration. At the 12.70 g/L copper containing solution the current efficiency was found to be 38.95% in 7h duration and the corresponding power consumption was 12.12 kWh/kg copper powder. Purity of copper was 99.9%.

37.4.3 *Reduction of Ammonium Paratungstate (APT)*

The detailed studies on tungsten powder reduction and kinetic investigation were carried out by the same author, Kumar et al. [17]. In this investigation the bed depth of APT was increased from 1.3 to 4 mm at the temperature of 1000 °C with time variation from 3 to 5 h time at a hydrogen flow rate of 0.1 LPM. The reduction percentage decreases from 99.99 to 79.51% with increase in bed depth from 1.3 to

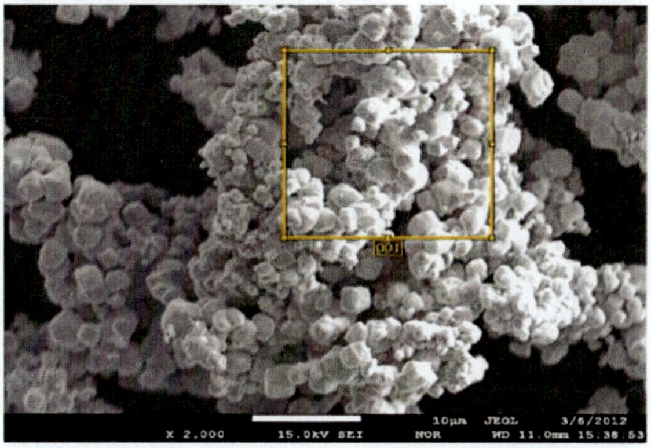

Fig. 37.2 Electron probe analysis of reduced tungsten powder

Table 37.3 Rate of reduction of APT with time and bed depth

APT (g)	Bed-depth (mm)	Time (h)	Reduction (%)
2	1.3	3	99.99
30	4	3	79.51
30	4	4	90.97
30	4	5	90.97

4mm in a time period of 180 min. This can be attributed to the less contact surface area of APT exposed to the hydrogen gas passing through. With the increase in time from 3 to 4h, the increase in reduction percentage was from 79.91 to 90.97%, beyond which no increase in reduction with time was found. The Electron Probe Micro Analysis (EPMA) of tungsten powder is given in Fig. 37.2, in which copper and cobalt were not detected. The purity of tungsten powder was 99.8% (Table 37.3).

37.5 Conclusions

1. A leaching-solvent extraction-electrowinning-reduction process is developed to recover the tungsten and copper from tungsten-copper (W-Cu) borings as tungsten powder and copper powder.
2. Leaching process parameters is optimized at 500 g scale. It has been found that 95.0% dissolution of W-Cu borings can be dissolved in 125 g/L ammonia at 1500 rpm stirring speed, 125 g/L solid: liquid ratio in 7h time.
3. The copper containing solution has been separated using the organic solvent 25% LIX84. The subsequent 20vol. % sulfuric acid used for stripping of copper. Electrowinning of copper has been done from the stripped solution containing 12.70 g/L copper at a better current efficiency of 38.95% with cathode current density of 725 A/m^2. The purity of copper was found to be 99.9%.

4. The recovery of tungsten from the separated tungsten containing ammonia solution was done by crystallization in a Rotary Vacuum Evaporator as ammonium paratungstate (APT).
5. APT was reduced upto 99.99% to high purity tungsten powder of 99.8%W at optimum process parameters of 1000 ^{0}C temperature, 0.1 LPM hydrogen flow rate and 3 h time. With increase in bed depth from 1.3 mm to 4 mm the reduction decreases from 99.99% to 90.97% in 4 h time.

Acknowledgement The authors are thankful to the Director, CSIR-NML, for giving permission to publish this paper.

References

1. Shen, L., Li, X., Lindberg, D., Taskinen, P.: Tungsten extractive metallurgy: a review of processes and their challenges for sustainability. Miner. Eng. (142), 105934 (2019).
2. U.S. Geological Survey: Mineral. Commodity Summaries 2021, pp. 178–179. USGS, Reston, Virginia, USA (2021).
3. Calvo, G., Valero, A., Valero, A.: How can strategic metals drive the economy? Tungsten and tin production in Spain during periods of war. Extr. Ind. Soc. **6**, 8–14 (2019)
4. Shedd, K. B.: Tungsten recycling in the United States in 2000, pp. 1-18. US Geological Survey, Reston, VA, USA (2005).
5. Luo, L., Kejun, L., Shibayama, A., Yen, W., Fujita, T., Shindo, O., Katai, A.: Recovery of tungsten and vanadium from tungsten alloy scrap. Hydrometall. (72), 1–8 (2004).
6. Kamal, S.S.K., Sahoo, P.K., Vimala, J., Shanker, B., Ghosal, P., Durai, L.: Synthesis of high purity tungsten nanoparticles from tungsten heavy alloy scrap by selective precipitation and reduction route. J. Alloys Compd. **678**, 403–409 (2016)
7. Venkateswaran, S., Schubert, W., Lux, B.: W-scrap recycling by the melt bath technique. Int. J. Refract. Metals Hard Mater. **4**(14), 263−270 (1996).
8. Bernhard, F. K., Lassner, E.: Reclamation of tungsten containing scrap materials: economic and technical aspects, Tungsten: 1987. In: Proceedings of the Fourth International Tungsten Symposium, Vancouver, 7-10 September, 59 – 67. MPR Publishing Services Ltd., Bellstone, Shrewsbury SY1 IHU, England (1987).
9. Lassner, E.: Modern methods of APT processing. Proceed. 2nd Int. Tungsten Symposium, San Francisco. 71–80. Mining Journals Books Ltd., 15, Wilson Street, London EC2M2TR, England (1982).
10. Lassner, E.: From Tungsten concentrates and scrap to highly pure ammonium paratungstate (APT). Int. J. Refract. Metals Hard Mater. (13), 35−44 (1995).
11. Van Put, J. W, Zegers, T., Van Sandwijk, W A., van der Straten, P. J. M.: proceedings of the 2nd International Conference on Separation Technology. In: Baird, M. H. I., Vijayan, C. (eds.) Hamilton, Ontario, Canada, pp 387−394. Canadian Society of Chemical Engineering, Canadian Institute of Mining and Metallurgy (1989).
12. Vadashdi, K.: Effluent-free manufacture of ammonia paratungstate (APT) by recycling the byproducts. Int. J. Refract. Metals Hard Mater. (3), 45−59 (1995).
13. D Coolidge, W.: Tungsten and method of making the same for use as filaments of incandescent electric lamps and for other purposes. United States Patent Office, Patent No. 1082933 (1913).
14. Fouad, N. E, Attyia, K. M. E., Zaki, M. I: Thermogravimetry of WO_3 reduction in hydrogen: kinetic charecterization of autocatalytic effects. Powder Technol. (74), 31−37 (1993).

15. Lackner, A., Filzwieser, A., Paschen, P.: On the reduction of tungsten blue oxide in a stream of hydrogen. Int. J. Refract. Metals Hard Mater. (14), 383–391 (1996).
16. Vahid, R., Seyed M. J., Reza, K. M., Reza, T. M. M.: Ammonia leaching: a new approach of copper industry in hydrometallurgical processes. J. Inst. Eng. (India): Ser. D. **94**(2), 95–104 (2013).
17. Kumar, M., Sau, D.C., Nishant, P., Bhattacharya, M.: Reduction of ammonium paratungstate generated during hydrometallurgical processing of tungsten-copper borings. Adv. Mater. Res. **828**, 123–134 (2014)